Engineer-In-Training
Exam File

Eleventh Edition

Donald G. Newnan, Ph.D., P.E.
Dean Newnan, Ph.D.

ENGINEERING PRESS, INC.   **SAN JOSE, CALIFORNIA**

**The first nine editions were titled
ENGINEER-IN-TRAINING LICENSE REVIEW**

Printed in the United States of America
5   4   3   2   1

Library of Congress Cataloging-in-Publication Data

Newnan, Donald G.
    Engineer-in-training exam file / Donald G. Newnan, Dean Newnan. --
11th ed.
        p.   cm. -- (Exam file series)
    Previous ed. C. Dean Newnan's name appeared first on t.p.
    ISBN 0-910554-78-1
    1.   Engineering--Examinations, questions, etc.   2. Engineering-
-Problems, exercises, etc.   I. Newnan, C. Dean.   II. Title.
III. Series.
TA159.N415   1990
620'.0076--dc20   89-23413
                                                         CIP

Engineering Press, Inc.   P.O. Box 1   San Jose, California 95103-0001

Contents

Exam Files

Professors around the country have opened their exam files and revealed their examination problems and solutions. These are actual exam problems with the complete solutions prepared by the same professors who wrote the problems. EXAM FILES are currently available for these topics:

Calculus I
Calculus II
Calculus III
Chemistry
Circuit Analysis
College Algebra
Differential Equations
Dynamics
Engineering Economy
Fluid Mechanics
Linear Algebra
Materials Science
Mechanics of Materials
Organic Chemistry
Physics I Mechanics
Physics II Heat, Light and Sound
Physics III Electricity and Magnetism
Probability and Statistics
Statics
Thermodynamics

The EXAM FILE series also includes four engineering license review books:

Engineer-In-Training Exam File
Civil Engineering License Exam File
Electrical Engineering License Exam File
Mechanical Engineering License Exam File

For a description of all available EXAM FILES, or to order them, ask at your college or technical bookstore, or write to:

Engineering Press, Inc.
P.O. Box 1
San Jose, California 95103-0001

Preface

Registration as a Professional Engineer is achieved by passing an Engineering Fundamentals (Engineer-In-Training) examination, followed by an exam in a specific branch of engineering. The first examination, whether called Engineer-In-Training, Intern Engineer, or Engineering Fundamentals, is a test of the applicant's knowledge of fundamentals of engineering and science. About 45,000 people take this exam each year and about 65% pass it. The second or professional engineering examination is a test of the applicant's ability to apply engineering principles and judgement to professional problems. Here we are concerned exclusively with the Engineer-In-Training or Engineering Fundamentals examination. A single national exam, prepared by the National Council of Engineering Examiners, is used in all 50 states.

The examination consists of two four-hour sessions. The morning session contains 140 multiple choice questions. There are five choices from which to select the correct answer for each question. The questions are bunched by topic. All 140 questions are required. The topics and approximate number of questions are:

Mathematics	17
Statics	12
Dynamics	17
Mechanics of Materials	12
Materials Science	6
Fluid Mechanics	14
Thermodynamics	14
Electrical Circuits	18
Engineering Economics	6
Chemistry	10
Structure of Matter	6
Computer Programming	8
	140

One must carefully pace their time in the morning, for an average of only 1.7 minutes per question may permit understanding the questions, but allows very little time for computation, and permits essentially no time to consult reference materials. One strategy used by many applicants is to rapidly work through the entire group of questions, answering the easy, fast ones, then to return to the more demanding questions in areas of personal competence. Finally, to guess as necessary to complete the questions.

In the afternoon four-hour session applicants are required to solve 50 questions as follows:

Engineering Mechanics (Statics, Dynamics, and Mechanics of Materials)	15
Mathematics	15
Electrical Circuits	10
Engineering Economics	10
	50

In addition, each applicant is required to solve 20 problems in two of five additional subject areas:

Computer Programming	10
Electronics and Electrical Machinery	10
Fluid Mechanics	10
Mechanics of Materials	10
Thermodynamics/Heat Transfer	10
	50

Thus the applicant must do 70 problems in the afternoon. The problems will require the application of engineering principles to slightly more complex problems in the afternoon than in the morning.

The test is open book, but there may be variation among the states (the State Boards are listed in the Appendix) regarding what may be brought into the examination. The exam is machine scored on the basis of correct answers only, so guessing is desirable. The afternoon problems have twice the grading weight of the morning problems, making a possible raw score of 280. The usual passing score of 70 is assigned to an approximate raw score of 140. This means one must answer about half the exam correctly to obtain a passing score.

The over 500 problems in the book have been selected to offer guidance in your review of the 13 subject areas. In this edition the book has been enlarged to add additional computer programming problems, an expanded engineering economics chapter, a detailed listing of the content of the Engineer-In-Training exam topics (in an Appendix), and a set of compound interest tables. I have appreciated past reader comments, and would appreciate being told about any errors that are noted. Please write me at the Engineering Press address. Finally, good luck in the exam!

DGN

MATHEMATICS

MATH 1

The integral of $y = x^3 - x + 1$ is:

$$\text{(a)} \quad 3x^2 - 1 + C$$

$$\text{(b)} \quad \frac{x^3}{3} - \frac{x^2}{2} + x$$

$$\text{(c)} \quad \frac{x^4}{3} - \frac{x^2}{2} + 1 + C$$

$$\text{(d)} \quad \frac{x^4}{3} - \frac{x^2}{2} + 1$$

$$\text{(e)} \quad \frac{x^4}{4} - \frac{x^2}{2} + x + C$$

$$\int y\,dx = \int (x^3 - x + 1)dx = \frac{x^4}{4} - \frac{x^2}{2} + x + C \quad \bullet$$

Answer is (e)

MATH 2

The only point of inflection on the curve representing the equation
$y = x^3 + x^2 - 3$ is at x equals:

$$\text{(a)} \quad -2/3$$
$$\text{(b)} \quad -1/3$$
$$\text{(c)} \quad 0$$
$$\text{(d)} \quad 1/3$$
$$\text{(e)} \quad 2/3$$

Maxima, minima and points of inflection:

Take first derivative, equate it to zero, and solve for its roots.

Take second derivative, insert the above roots to determine whether
each is a maximum or a minimum.

Equate second derivative to zero and solve for its root(s) or point(s)
of inflection.

	Max.	Min.	Inflection
$\dfrac{dy}{dx}$	0	0	anything
$\dfrac{d^2y}{dx^2}$	-	+	0

Apply the above to the stated problem: $y = x^3 + x^2 - 3$

$\dfrac{dy}{dx} = 3x^2 + 2x = 0 \qquad x(3x + 2) = 0 \qquad \therefore x = 0, -2/3$
(max. and min. exist here)

$\dfrac{d^2y}{dx^2} = 6x + 2 \qquad$ at $x = 0$: $\dfrac{d^2y}{dx^2} = +2$, minimum exists

at $x = -2/3$: $\dfrac{d^2y}{dx^2} = -4 + 2 = -2$, maximum exists

$\dfrac{d^2y}{dx^2} = 6x + 2 = 0 \qquad x = -1/3$, inflection point exists ●

Answer is (b)

MATH 3

There are 10 defectives per 1,000 items of a product in the long run. What is the probability that there is one and only one defective in a random lot of 100?

(a) 99×0.01^{99}

(b) 0.01

(c) 0.5

(d) 0.99^{100}

(e) 0.99^{99}

The problem involves the binomial probability.

$P_{defective} = \dfrac{10}{1000} = 0.01 \qquad$ This is the probability that one item selected at random is defective.

$P_{good} = 1 - P_{defective} = 0.99$

The probability that exactly one defective will be found in a random sample of 100 items is given by the binomial: $b(1, 100, 0.01)$.

$b(1, 100, 0.01) = {}_{100}C_1 \, (0.01)^1 \, (0.99)^{99}$

$\underbrace{\quad}_{P_{defective}} \qquad \underbrace{\quad}_{P_{good}}$

$_nC_r = \begin{pmatrix} n \\ r \end{pmatrix} = \dfrac{n!}{(n-r)! \; r!}$ = number of combinations of n objects taken r at a time without concern for order of arrangement.

$_{100}C_1 = \dfrac{100!}{99! \; 1!} = 100 \qquad b(1, 100, 0.01) = 100(0.01)(0.99)^{99} = 0.99^{99}$ ●
$= 0.3697$

Answer is (e)

MATH 4

If the second derivative of the equation of a curve is equal to the negative of the equation of that same curve, the curve is

(a)   an exponential
(b)   a tangent
(c)   a conic section
(d)   a sinusoid
(e)   a parabola

Taking the sine curve
$$f(x) = \sin x \qquad f'(x) = \cos x \qquad f''(x) = -\sin x$$

Thus a function whose second derivative is equal to the negative of the equation of that same function is a sinusoid. ●
Answer is (d)

MATH 5

To find the angles of a triangle, given only the lengths of the sides, one would use

(a)   the law of cosines
(b)   the law of tangents
(c)   the law of sines
(d)   the inverse-square law
(e)   orthogonal functions

The law of cosines is
$$a^2 = b^2 + c^2 - 2bc \, \cos A$$

for any plane triangle with angles A, B, C and sides a, b, c respectively.

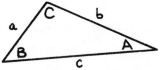

This law can be applied to solve for the angles, given three sides in a plane triangle.

Answer is (a) ●

MATH 6

The rationalized value of the complex number $\dfrac{6 + i\,2.5}{3 + i\,4}$ is

(a) $2 + i\,0.625$

(b) $\left(\dfrac{6 + i\,2.5}{3 + i\,4}\right)\left(\dfrac{6 - i\,2.5}{3 + i\,4}\right)$

(c) $\left(\dfrac{6 + i\,2.5}{3 + i\,4}\right)\left(\dfrac{6 + i\,2.5}{3 + i\,4}\right)$

(d) $\dfrac{28 - i\,16.5}{25}$

(e) $1 - i$

To rationalize, multiply the numerator and denominator of the complex fraction by the complex conjugate of the denominator.

$$\frac{6 + i\,2.5}{3 + i\,4} \times \frac{3 - i\,4}{3 - i\,4} = \frac{18 - i\,24 + i\,7.5 + 10}{9 - i\,12 + i\,12 + 16} = \frac{28 - i\,16.5}{25} \quad ●$$

Answer is (d)

MATH 7

Study the series of numbers to discover the "SYSTEM" in which they are arranged.

For the series   1   5   14   30   __   91   the fifth term is

      (a)   59
      (b)   53
      (c)   61
      (d)   50
      (e)   55

If we are given any series of values it is always possible to find a polynomial which will pass through the given points. So the equation of the polynomial would be the "system." Thus <u>any</u> number between 30 and 91 could be used.

$$
\begin{array}{ccccccc}
& 1 & 5 & 14 & 30 & x & 91 \\
\text{first differences:} & 4 & 9 & 16 & y & z \\
& 2^2 & 3^2 & 4^2 & 5^2 & 6^2
\end{array}
$$

We see that the first differences are successive perfect squares.

    Thus $x - 30 = y = 5^2 = 25$ $x = 30 + 25 = 55$ ●

                  Answer is (e)

MATH 8

If $\sin \alpha = \dfrac{a}{\sqrt{a^2 + b^2}}$ which of the following statements is true?

      (a) $\tan^{-1} \dfrac{b}{a} = \dfrac{\pi}{2} - \alpha$

      (b) $\tan^{-1} \dfrac{b}{a} = - \alpha$

      (c) $\cos^{-1} \dfrac{b}{\sqrt{a^2 + b^2}} = \dfrac{\pi}{2} - \alpha$

      (d) $\cos^{-1} \dfrac{a}{\sqrt{a^2 + b^2}} = \alpha$

      (e) None of the above

So the triangle looks like this:

$\tan \left(\dfrac{\pi}{2} - \alpha \right) = \dfrac{b}{a}$

$\therefore \tan^{-1} \dfrac{b}{a} = \dfrac{\pi}{2} - \alpha$ ●

          Answer is (a)

MATH 9

In finding the distance between two points P_1 (X_1, Y_1) and P_2 (X_2, Y_2), the most direct procedure is to use:

    (a)   The translation of the axes
    (b)   The Pythagorean Theorem
    (c)   The slope of the line
    (d)   The derivative
    (e)   The integral

The Pythagorean Theorem

$$d^2 = (X_2 - X_1)^2 + (Y_2 - Y_1)^2$$

Answer is (b) ●

MATH 10

The sine of $840°$ equals:

    (a) $-\cos 30°$
    (b) $-\cos 60°$
    (c) $\sin 30°$
    (d) $\sin 60°$
    (e) $-\sin 30°$

$840° = 2(360) + 120 = 2\pi + 120°$

$\sin (2\pi + 120°) = \sin 120° = \sin (180° - 60°) = \sin 60°$ ●

Answer is (d)

MATH 11

One root of $x^3 - 8x - 3 = 0$ is:

    (a) 2
    (b) 3
    (c) 4
    (d) 5
    (e) 6

The solution is obtained by seeing which of the 5 answers satisfies the equation.

x	x^3-8x-3
2	-11
3	0 ●
4	29
5	82
6	165

Answer is (b)

MATH 12

The value of Tan (A + B), where Tan A = $\frac{1}{3}$ and Tan B = $\frac{1}{4}$
is   (Note: A and B are acute angles)

    (a) 7/12
    (b) 1/11
    (c) 7/11
    (d) 7/13
    (e) none of the above

Sin(A + B) = Sin A Cos B + Cos A Sin B

Cos(A + B) = Cos A Cos B - Sin A Sin B

So Tan(A + B) = $\frac{Sin(A + B)}{Cos(A + B)}$ = $\frac{Sin\ A\ Cos\ B + Cos\ A\ Sin\ B}{Cos\ A\ Cos\ B - Sin\ A\ Sin\ B}$

dividing by Cos A Cos B

$$Tan(A + B) = \frac{\dfrac{Sin\ A\ Cos\ B}{Cos\ A\ Cos\ B} + \dfrac{Cos\ A\ Sin\ B}{Cos\ A\ Cos\ B}}{\dfrac{Cos\ A\ Cos\ B}{Cos\ A\ Cos\ B} - \dfrac{Sin\ A\ Sin\ B}{Cos\ A\ Cos\ B}}$$

$$= \frac{Tan\ A + Tan\ B}{1 - Tan\ A\ Tan\ B}$$

$$= \frac{\dfrac{1}{3} + \dfrac{1}{4}}{1 - \dfrac{1}{3} \times \dfrac{1}{4}} = \frac{\dfrac{4}{12} + \dfrac{3}{12}}{1 - \dfrac{1}{12}} = \frac{\dfrac{7}{12}}{\dfrac{11}{12}} = \frac{7}{11} \ ●$$

Note that the problem could also be solved by determining Angle A
(whose tangent is 1/3) and Angle B (whose tangent is 1/4). Then we
could find the tangent of (A + B).

$Tan^{-1} \frac{1}{3}$ = 18.435° $Tan^{-1} \frac{1}{4}$ = 14.036°

Tan(18.435° + 14.036°) = Tan(32.471°) = 0.6364 = $\frac{7}{11}$ ●

Answer is (c)

MATH 13

To cut a right circular cone in such a was as to reveal a parabola, it
must be cut

    (a) perpendicular to the axis of symmetry
    (b) at any acute angle to the axis of symmetry
    (c) at any obtuse angle to the axis of symmetry
    (d) parallel to the axis of symmetry
    (e) none of these

To reveal a parabola, a right circular cone must be cut parallel to an
element of the cone and intersecting the axis of symmetry. ●

Answer is (e)

MATH 14

Naperian logarithms have a base of:

            (a)   3.1416
            (b)   2.171828
            (c)   10
            (d)   Any positive number can be used
            (e)   2.71828

Common logarithms have base 10.
Naperian or natural logarithms have base e = 2.71828 ●

Answer is (e)

MATH 15

$(5.743)^{1/30}$ equals:

            (a)   1.03
            (b)   1.04
            (c)   1.05
            (d)   1.06
            (e)   1.07

Log $(5.743)^{1/30}$ = $\frac{1}{30}$ Log 5.743 = $\frac{1}{30}$ (0.7592) = 0.0253

The antilogarithm of 0.0253 is 1.06 ●

Answer is (d)

MATH 16

For a given curve f(x,y) = 0, it is found that f(x,y) = f(x,-y).
This means that

            (a)   The curve is unsymmetrical
            (b)   The curve is a conic section
            (c)   The curve is symmetrical to both axes
            (d)   The curve is symmetrical to the x-axis
            (e)   The curve is symmetrical to the y-axis

Answer is (d) ●

MATH 17

The curve in the figure below has the equation y = f(x). At point A, what
are the values of $\frac{dy}{dx}$ and $\frac{d^2y}{dx^2}$?

(a) $\dfrac{dy}{dx} < 0$, $\quad \dfrac{d^2y}{dx^2} < 0$ $\qquad$ (d) $\dfrac{dy}{dx} > 0$, $\quad \dfrac{d^2y}{dx^2} > 0$

(b) $\dfrac{dy}{dx} < 0$, $\quad \dfrac{d^2y}{dx^2} > 0$ $\qquad$ (e) $\dfrac{dy}{dx} > 0$, $\quad \dfrac{d^2y}{dx^2} < 0$

(c) $\dfrac{dy}{dx} = 0$, $\quad \dfrac{d^2y}{dx^2} = 0$

The first derivative $\dfrac{dy}{dx}$ is the slope of the curve. At point A the slope is positive. The second derivative $\dfrac{d^2y}{dx^2}$ gives the direction of bending. A negative value indicates the curve is concave downward.

Answer is (e) ●

MATH 18

The expression $\quad \dfrac{6!}{3! \; 0!} \quad$ is equal to

(a) ∞
(b) 120
(c) 2!
(d) 0
(e) 3!

$$\dfrac{6!}{3! \; 0!} = \dfrac{6 \times 5 \times 4 \times 3!}{3! \times 1} = 120 \; ●$$

Answer is (b)

MATH 19

$$\lim_{x \to 1} \dfrac{x^2 - 1}{x - 1} =$$

(a) 2
(b) ∞
(c) 0
(d) 1
(e) none of these

$$\lim_{x \to 1} \dfrac{x^2 - 1}{x - 1} = \lim_{x \to 1} \dfrac{(x - 1)(x + 1)}{(x - 1)} = \lim_{x \to 1} (x + 1) = 2 \; ●$$

Answer is (a)

MATH 20

In statistics, the standard deviation measures
(a) a standard distance
(b) a normal distance
(c) central tendency
(d) dispersion
(e) a unit distance

Answer is (d) ●

MATH 21

There are 3 bins containing integrated circuits. One bin has 2 premium IC's, one has 2 regular IC's, and one has 1 premium IC and 1 regular IC.

You choose an IC at random without looking into any of the bins. The chosen IC is found to be a premium IC.

What is the probability that the remaining IC in that bin is also a premium IC?

(a) 1/5
(b) 1/4
(c) 1/3
(d) 2/3
(e) 1/2

Since the first IC chosen is a premium IC, it was drawn from either Bin 1 or Bin 3. From the distribution of premium IC's, the probability that the premium IC came from Bin 1 is 2/3, and from Bin 3 is 1/3.

In Bin 1 the probability that the remaining IC is a premium IC is 1; in Bin 3 the probability is 0. Thus the probability that the remaining IC is a premium IC is:

$$\frac{2}{3}(1) + \frac{1}{3}(0) = \frac{2}{3} \; \bullet$$

Alternate solution using Bayes' Theorem for conditional probability:

$$P\left(\text{Bin 1} \mid \text{Drew Premium}\right) = \frac{P(\text{Bin 1 and Premium})}{P(\text{Premium})} = \frac{P\left(\text{Premium} \mid \text{Bin 1}\right) \cdot P(\text{Bin 1})}{\sum_{i=1}^{3} P\left(\text{Premium} \mid \text{Bin 1}\right) \cdot P(\text{Bin 1})}$$

$$= \frac{1\left(\frac{1}{3}\right)}{1\left(\frac{1}{3}\right) + 0\left(\frac{1}{3}\right) + \frac{1}{2}\left(\frac{1}{3}\right)} = \frac{2}{3} \; \bullet$$

Answer is (d)

MATH 22

If a right circular cone is cut parallel with the axis of symmetry, you would reveal

(a) circle
(b) hyperbola
(c) eclipse
(d) parabola
(e) none of these

Answer is (b) ●

MATH 23

The simplest value of $\dfrac{\left[(n+1)!\right]^2}{n!\,(n-1)!}$ is

(a) n^2

(b) $n(n+1)$

(c) $n+1$

(d) $n(n+1)^2$

(e) $\dfrac{n+1}{n-1}$

The value $(n+1)!$ may be written as $(n+1)(n)\left[(n-1)!\right]$

It may also be written as $n!\,(n+1)$

Hence the given expression may be written as follows:

$$\frac{\left\{(n+1)(n)\ \left[(n-1)!\right]\right\}\left\{n!\,(n+1)\right\}}{n!\,(n-1)!} = (n+1)^2 n \;\bullet$$

Answer is (d)

MATH 24

If $X^{3/4} = 8$, X equals

(a) 6

(b) 9

(c) -9

(d) 16

(e) 20

Raise both sides of the equation to the 4/3 power

$$\left[X^{3/4}\right]^{4/3} = 8^{4/3}$$

$$X = \sqrt[3]{8^4} = \sqrt[3]{(2^3)^4} = 2^{\frac{3\cdot4}{3}} = 2^4 = 16 \;\bullet$$

Answer is (d)

MATH 25

If $\log_a 10 = 0.250$, $\log_{10} a$ equals

(a) 4

(b) .50

(c) 2

(d) .25

(e) 1,000

$\log_a 10 = 0.250$ can be written as $10 = a^{0.250}$

taking $\log_{10}$ $\log_{10} 10 = \log_{10} a^{0.250}$

$$1 = 0.250 \log_{10} a$$

Since $1 = 0.250 \log_{10} a$

$$\log_{10} a = \frac{1}{0.250} = 4 \quad \bullet$$

Answer is (a)

MATH 26

The csc of $960°$ is equal to

(a) $-\frac{2\sqrt{3}}{3}$

(b) 1

(c) 1/2

(d) -2

(e) $-\frac{\sqrt{3}}{2}$

$$\csc 960° = \csc(5\pi + 60)$$
$$= \csc \theta = \frac{r}{y} = \frac{2}{-\sqrt{3}} = -\frac{2\sqrt{3}}{3} \quad \bullet$$

Answer is (a)

MATH 27

If $i = \sqrt{-1}$, the quantity i^{27} is equal to

(a) 0

(b) i

(c) -i

(d) 1

(e) -1

$$i^{27} = (\sqrt{-1})^{27} = \sqrt{(-1)^{27}} = \sqrt{(-1)^{26}(-1)} = (-1)^{13}\sqrt{-1} = -1\sqrt{-1}$$

now substituting back i for $\sqrt{-1}$

$$i^{27} = -1(i) = -i \quad \bullet$$

Answer is (c)

MATH 28

If the first derivative of the equation of a curve is a constant, the curve is a

(a) circle

(b) hyperbola

(c) parabola

(d) straight line

(e) sine wave

If $\frac{dy}{dx} = m, \quad y = \int m \, dx = m \int dx = mx + b$

so $y = mx + b$ a straight line $\bullet$

Answer is (d)

MATH 29

An angle between 90° and 180° has:

      (a) A positive sine and cosine
      (b) A negative cotangent and cosecant
      (c) A negative secant and tangent
      (d) All its trigonometric functions negative
      (e) All its trigonometric functions positive

In the second quadrant all functions
are negative except the sine and
cosecant. ●

Answer is (c)

MATH 30

The following is a characteristic of all trigonometric functions:

      (a) The values of all functions repeat themselves every 45 degrees.
      (b) All functions are units of length or angular measure.
      (c) The graphs of all functions are continuous.
      (d) The values of all functions are never greater than 1.00.
      (e) All functions are dimensionless units.

All functions are ratios of lengths with the result that they are
dimensionless. ●

Answer is (e)

MATH 31

Given the function $P = 2R^2S^3T^{1/2} + R^{1/3}S \sin 2T$
$\frac{\partial P}{\partial T}$ is

      (a) $R^2S^3T^{1/2} + 2R^{1/3}S \cos 2T$

      (b) $R^2S^3T^{-1/2} + 2R^{1/3}S \cos 2T$

      (c) $2R^2S^3T^{-1/2} + R^{1/3}S \cos 2T$

      (d) $R^2S^3T^{-1/2} + 2R^{1/3}S \sin 2T$

      (e) $2R^2S^3T^{1/2} + R^{1/3}S \sin 2T$

In finding $\frac{\partial P}{\partial T}$ R and S are treated as constants.

$$P = 2R^2S^3T^{1/2} + R^{1/3}S \sin 2T$$

$$\frac{\partial P}{\partial T} = 2R^2S^3(\frac{1}{2} T^{-1/2}) + R^{1/3}S(\cos 2T)(2)$$

$$= R^2S^3T^{-1/2} + 2R^{1/3}S \cos 2T ●$$

Answer is (b)

MATH 32

For a given curve $y = f(x)$ that is continuous between $x = a$ and $x = b$, the average value of the curve between the ordinates at $x = a$ and $x = b$ is represented by:

(a) $\dfrac{\displaystyle\int_a^b x^2\, dy}{(b-a)}$

(b) $\dfrac{\displaystyle\int_a^b y^2\, dx}{(b-a)}$

(c) $\dfrac{\displaystyle\int_a^b x\, dy}{(a-b)}$

(d) $\dfrac{\displaystyle\int_a^b y\, dx}{(a-b)}$

(e) $\dfrac{\displaystyle\int_a^b y\, dx}{(b-a)}$

$$\text{Area} = \int_a^b y\, dx$$

$$\text{Average value} = \frac{\text{area}}{\text{base width}} = \frac{\displaystyle\int_a^b y\, dx}{b-a} \quad \bullet$$

Answer is (e)

MATH 33

The probability that both stages of a 2-stage missile will function correctly is 0.95. The probability that the first stage will function correctly is 0.98.

What is the probability that the second stage will function correctly given that the first one does?

(a) 0.99
(b) 0.98
(c) 0.97
(d) 0.95
(e) 0.93

Given: $P(S_1) = 0.98$ $\qquad P(S_2 \cap S_1) = 0.95$ $\qquad$ Find: $P(S_2 | S_1)$

$$P(S_2 | S_1) = \frac{P(S_2 \cap S_1)}{P(S_1)} = \frac{0.95}{0.98} = 0.97 \quad \bullet$$

Answer is (c)

MATH 34

If $y = \cos x$, $\dfrac{dy}{dx} = ?$

(a) $\sin x$

(b) $-\tan x \cos x$

(c) $\dfrac{1}{\sec x}$

(d) $\sec x \sin x$

(e) $-\dfrac{1}{\cos x}$

$$\frac{dy}{dx} = -\sin x$$ Since $\tan x = \frac{\sin x}{\cos x}$,   then $\sin x = \tan x \cos x$

Thus we can define the derivative

$$\frac{dy}{dx} = -\tan x \cos x \ \bullet$$

Answer is (b)

MATH 35

The value of the determinant $\begin{vmatrix} 1 & 1 & 0 & 0 \\ 1 & 1 & 1 & 0 \\ 0 & 1 & 1 & 1 \\ 0 & 0 & 1 & 1 \end{vmatrix}$ is

(a) +1
(b)   0
(c) -2
(d) -1
(e) none of the above values

$$\begin{vmatrix} 1 & 1 & 0 & 0 \\ 1 & 1 & 1 & 0 \\ 0 & 1 & 1 & 1 \\ 0 & 0 & 1 & 1 \end{vmatrix} = 1 \begin{vmatrix} 1 & 1 & 0 \\ 1 & 1 & 1 \\ 0 & 1 & 1 \end{vmatrix} - 1 \begin{vmatrix} 1 & 0 & 0 \\ 1 & 1 & 1 \\ 0 & 1 & 1 \end{vmatrix}$$

$$= 1 \begin{vmatrix} 1 & 1 \\ 1 & 1 \end{vmatrix} - 1 \begin{vmatrix} 1 & 0 \\ 1 & 1 \end{vmatrix} - 1 \begin{vmatrix} 1 & 1 \\ 1 & 1 \end{vmatrix} + 1 \begin{vmatrix} 0 & 0 \\ 1 & 1 \end{vmatrix} = -1(1) + 1(0) = -1 \ \bullet$$

Answer is (d)

MATH 36

The following equation involves two determinates.

$$\begin{vmatrix} 3 & X \\ 2 & 2 \end{vmatrix} = \begin{vmatrix} 2 & -1 \\ X & -3 \end{vmatrix}$$

The value of X is nearest to

(a) -1
(b)   1
(c)   2
(d)   3
(e)   4

$$6 - 2X = -6 - (-X)$$

$$6 - 2X = -6 + X$$ Collecting like terms gives $12 = 3X$

So $4 = X \ \bullet$

Answer is (e)

MATH 37

Let $x_1 = (2,-4,5,6)$.

The value of $\|x_1\|$ is closest to

(a)   2
(b)   6
(c)   9
(d)   65
(e)   81

$$\|x_1\| = \sqrt{2^2 + (-4)^2 + 5^2 + 6^2} = \sqrt{4 + 16 + 25 + 36} = \sqrt{81} = 9 \,\bullet$$

Answer is (c)

MATH 38

Given

$$\begin{bmatrix} X & 5 \\ 3 & X+Y \end{bmatrix} = \begin{bmatrix} 0 & 5 \\ 3 & 2 \end{bmatrix}$$

The value of Y is nearest to

(a)   0
(b)   2
(c)   3
(d)   4
(e)   5

If two matrices are equal, then their corresponding elements must be equal. Therefore $X = 0$, and $X+Y = 2$. Since $X = 0$, Y must equal 2. $\bullet$

Answer is (b)

MATH 39

Let $\vec{a} = (1,2,2)$ and $\vec{b} = (3,0,4)$. The standard inner product $\langle \vec{a}, \vec{b} \rangle$ of a and b is closest to

(a)   0
(b)   2
(c)   5
(d)   7
(e)   11

$$\langle \vec{a}, \vec{b} \rangle = 1(3) + 2(0) + 2(4) = 11 \,\bullet$$

Answer is (e)

MATH 40

Let $\vec{u} = \begin{bmatrix} 2 \\ 1 \\ 1 \end{bmatrix}$ $\vec{v} = \begin{bmatrix} 2 \\ t \\ 2t \end{bmatrix}$

The value of t for which $\vec{u}$ is orthogonal to $\vec{v}$ is closest to

(a)   −4/3
(b)   3/4
(c)   0
(d)   3
(e)   4

For $\vec{u} \perp \vec{v}$, we need $\vec{u}^T \cdot \vec{v} = 0$

$\vec{u}^T \cdot \vec{v} = 2(2) + 1(t) + 1(2t) = 4 + 3t = 0$

So t = −4/3 ●

Answer is (a)

MATH 41

A standard deck of 52 playing cards is well shuffled. The probability that the first four cards dealt from the deck will be the four aces is closest to

(a) 2.0×10^{-1}
(b) 8.0×10^{-2}
(c) 4.0×10^{-4}
(d) 2.0×10^{-6}
(e) 4.0×10^{-6}

The probability of drawing an ace on the first card is $\frac{4}{52}$

The probability the second card is an ace is $\frac{3}{51}$

The probability the third card is an ace $\frac{2}{50}$ and the fourth card $\frac{1}{49}$.

The probability the first four cards will be aces is

$P = \frac{4}{52} \cdot \frac{3}{51} \cdot \frac{2}{50} \cdot \frac{1}{49} = 0.0000037 = 3.7 \times 10^{-6}$ ●

Answer is (e)

STATICS 1

The magnitude of the force acting in the member BC of the truss shown below is:

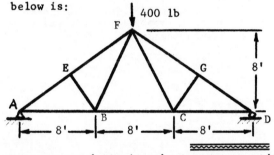

(a) 200 lb.
(b) 300 lb.
(c) 360 lb.
(d) 400 lb.
(e) 480 lb.

By symmetry determine the support reaction A_y to be 200 lb.

Although A is pinned, there are no horizontal forces on the truss, so $A_x = 0$.

Section the truss along ① to cut BC. Draw the free body diagram of the structure to the left of the section.

Take moments about F. (Counterclockwise is positive.)

$$M_F = 0 = -12(200) + 8(BC)$$

$$BC = +300 \text{ lb} \bullet$$
Tension

Answer is (b)

STATICS 2

The coefficient of friction between the 60-pound block and the plane shown is 0.30. If the block is to remain in equilibrium, what is the maximum allowable magnitude for the force P?

(a) 12 lb.
(b) 15 lb.
(c) 18 lb.
(d) 21 lb.
(e) 24 lb.

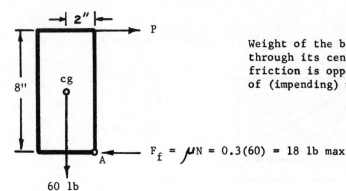

Weight of the block acts downward through its center of gravity and friction is opposite to direction of (impending) motion.

$F_f = \mu N = 0.3(60) = 18$ lb max

To avoid translation: $F_x = 0 = +P - 18_{max}$ $\therefore$ $P = + 18$ lb. max. ⟶

To avoid tipping about A:

$$M_A = 0 = -8P + 2(60) \quad \therefore \quad P = + 15 \text{ lb. max.} \longrightarrow$$

Choose the lesser value of P, or 15 lb. ●

Answer is (b)

STATICS 3

For a system to be in equilibrium, the sum of the external forces acting on the system must be

(a) equal to unity
(b) a maximum
(c) indeterminant
(d) zero
(e) infinite

Conditions of equilibrium for a system are:

1. The sum of the external forces equal zero
2. The sum of the external moments equal zero

Answer is (d)●

STATICS 4

Derive the formula for computing the moment of interia of an equilateral triangle about one of its sides.

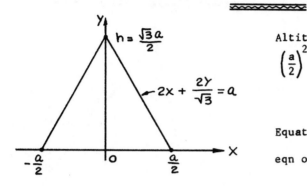

Altitude h

$$\left(\frac{a}{2}\right)^2 + h^2 = a^2 \qquad h^2 = a^2 - \frac{a^2}{4} = \frac{3a^2}{4}$$

$$h = \frac{\sqrt{3}\,a}{2}$$

Equation of side in terms of x and y

eqn of straight line (intercept form)

$$\frac{x}{\frac{a}{2}} + \frac{y}{\frac{\sqrt{3}\,a}{2}} = 1$$

$$2x + \frac{2y}{\sqrt{3}} = a \qquad 2x = a - \frac{2y}{\sqrt{3}}$$

Moment of Interia

$$I_x = \int y^2\, dA = \int y^2\left(a - \frac{2}{\sqrt{3}}\,y\right) dy = \int_{0}^{\frac{\sqrt{3}a}{2}} \left[ay^2 - \frac{2}{\sqrt{3}}\,y^3\right] dy$$

$$= \left[\frac{ay^3}{3} - \frac{2y^4}{4\sqrt{3}}\right]_{0}^{\frac{\sqrt{3}a}{2}}$$

$$= \frac{\sqrt{3}\,a^4}{8} - \frac{3\sqrt{3}\,a^4}{32} = \frac{\sqrt{3}}{32}\,a^4 \quad \bullet$$

STATICS 5

A tripod whose legs are each 10 feet long supports a load of 1000 pounds.

The feet of the tripod are at the vertices of a horizontal isosceles triangle whose base is 12 feet and whose altitude is 8 feet.

Determine the total load in each leg.

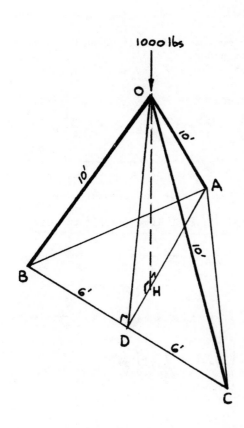

From the figure

$$OD = \sqrt{10^2 - 6^2} = \sqrt{64} = 8$$

Consider triangle ODA separately

$$\overline{OH}^2 + \overline{DH}^2 = 8^2 \qquad (1)$$

$$\overline{OH}^2 + (8 - DH)^2 = 10^2 \quad (2)$$
$$\underline{-\overline{OH}^2 - \overline{DH}^2 \qquad\quad = -8^2} \quad -(1)$$
$$(8 - DH)^2 - \overline{DH}^2 = 36$$

$$64 - 16\ DH + \overline{DH}^2 - \overline{DH}^2 = 36$$

$$DH = \frac{64 - 36}{16} = 1.75 \text{ ft.}$$

$$HA = 8 - 1.75 = 6.25 \text{ ft.}$$

Solving for the distance OH

$$\overline{OH}^2 + \overline{DH}^2 = \overline{OD}^2$$

$$\overline{OH}^2 + 1.75^2 = 8^2$$

$$\overline{OH}^2 = 64 - 3.06 = 60.94$$

$$OH = 7.8 \text{ ft.}$$

With the dimensions calculated we can now solve triangle ODA, assuming that members OB and OC have been replaced by the single member OD.

$$\Sigma F_x = 0$$

$$OD_x - OA_x = 0 \qquad \frac{1.75}{8}\ OD = \frac{6.25}{10}\ OA \qquad OA = \frac{10 \times 1.75}{8 \times 6.25}\ OD$$

$$OA = 0.35\ OD$$

$$\Sigma F_y = 0$$

$$OD_y + OA_y - 1000 = 0 \qquad \frac{7.8}{8}\ OD + \frac{7.8}{10}\ OA - 1000 = 0$$

$$0.976\ OD + 0.78(0.35\ OD) = 1000$$

$$0.976\ OD + 0.273\ OD = 1000$$

$$OD = \frac{1000}{1.249} = 800 \text{ lbs.}$$

$$OA = 0.35(800) = 280 \text{ lbs. } \bullet$$

Now we can replace member OD by members OB and OC

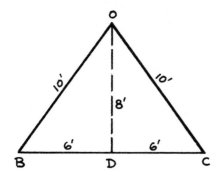

$\Sigma F_y = 0$

$OB_y + OC_y = OD$

Since $OB = OC$ and $OD = 800$ lbs.,

$2 \ OB_y = 800 \ ; \quad 2(\frac{8}{10}) \ OB = 800$

$OB = OC = \dfrac{800 \times 10}{2 \times 8} = 500 \text{ lbs. } \bullet$

(All legs are in compression.)

STATICS 6

Find the moment of inertia of the triangle shown below about its base.

By similar triangles $\dfrac{x}{4} = \dfrac{3 - y}{3}$

$x = \dfrac{4(3 - y)}{3} \qquad dA = xdy = \dfrac{4}{3}(3 - y)dy$

$I = \int y^2 dA = \dfrac{4}{3}\int_0^3 (3 - y)y^2 dy$

$= \dfrac{4}{3}\left[y^3 - \dfrac{y^4}{4} \right]_0^3 = \dfrac{4}{3}\left[27 - \dfrac{81}{4} \right]$

$= \dfrac{4}{3}(6.75) = 9 \text{ in}^4 \ \bullet$

STATICS 7

A weight W is suspended from the end of a 3-foot long weightless rod, as shown. Find an expression for the cosine of the angle θ when the system is in equilibrium. Neglect any friction between the contacting surfaces.

Since there is no friction $F_{A_y} = 0$

$\Sigma F_y = 0$ $F_{B_y} - W = 0$ $\therefore F_{B_y} = W$

$\Sigma F_x = 0$ $F_{A_x} - F_{B_x} = 0$ $\therefore F_{A_x} = F_{B_x}$

$\Sigma M_A = 0$ $(1 \times F_{B_y}) + (1 \times F_{B_x} \times \tan \theta) - (W \times 3 \times \cos \theta) = 0$

$$W + F_{B_x} \tan \theta - 3W \cos \theta = 0 \qquad (1)$$

$F_{B_x} = F_B \sin \theta$ $F_{B_y} = F_B \cos \theta$

$$\frac{F_{B_x}}{\sin \theta} = \frac{F_{B_y}}{\cos \theta} \qquad \therefore F_{B_x} = F_{B_y} \frac{\sin \theta}{\cos \theta} = W \frac{\sin \theta}{\cos \theta}$$

Substituting the value of F_{B_x} into (1)

$$W + W \frac{\sin \theta}{\cos \theta} \frac{\sin \theta}{\cos \theta} - 3W \cos \theta = 0$$

$$W \cos^2 \theta + W \sin^2 \theta - 3W \cos^3 \theta = 0$$

But $\cos^2 \theta + \sin^2 \theta = 1$

$$W - 3W \cos^3 \theta = 0 \qquad \cos^3 \theta = \frac{1}{3} \qquad \cos \theta = \sqrt[3]{\frac{1}{3}} \;\; \bullet$$

STATICS 8

The mass moment of inertia of a cylinder about its central axis perpendicular to a circular cross section is

(a) directly proportional to its radius
(b) independent of its radius
(c) directly proportional to its length
(d) independent of its length
(e) inversely proportional to the square of its radius

mass moment of inertia $= \int r^2 dM$

for a cylinder $dM = 2\pi r dr \cdot \rho L$

where r = radial distance to dr
ρ = density of cylinder
L = length of cylinder

$$I = \int_0^R r^2 2\pi r dr \cdot \rho L = 2\pi\rho L \int_0^R r^3 dr = 2\pi\rho L \left.\frac{r^4}{4}\right|_0^R = \frac{1}{2}\pi\rho LR^4$$

Thus the mass moment of inertia is proportional to the length of the cylinder. ●

Answer is (c)

STATICS 9

Which of the following is true with respect to the vector diagram shown below?

(a) $\vec{D} = \vec{A} + \vec{B} - \vec{C}$

(b) $2\vec{B} = \vec{A} + 2\vec{C}$

(c) $2\vec{A} = \vec{B} + \vec{D}$

(d) $\vec{C} = \vec{D} - \vec{A} - \vec{B}$

(e) $\vec{A} - \vec{B} = \vec{D} - \vec{C}$

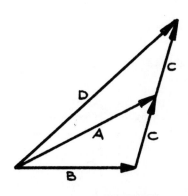

$$\vec{A} = \vec{D} - \vec{C}$$
$$\vec{A} = \vec{B} + \vec{C}$$
$$\overline{2\vec{A} = \vec{D} + \vec{B}} \quad \bullet$$

Answer is (c)

STATICS 10

The reaction at point "A" is

(a) zero

(b) 40 lbs ↑

(c) 40 lbs ↓

(d) 40 lbs ↑ plus 400 ft-lbs

(e) 40 lbs ↓ plus 400 ft-lbs

$\Sigma M_B = 0$

$10A + 100(0) = 0$

$\therefore \; A = 0 \quad \bullet$

Answer is (a)

STATICS 11

The beam shown at the right is
(a) stable
(b) unstable
(c) statically determinate
(d) statically indeterminate
(e) none of the above
     choices

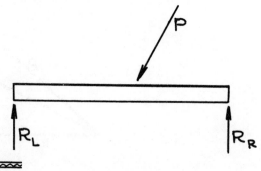

The load P has a horizontal component that is not resisted by the reactions R_L or R_R. Therefore the beam will slide and is thus unstable. ●
Answer is (b)

STATICS 12

An airtight closed box of weight P is suspended from a spring balance. A bird of weight W is placed on the floor of the box, and the balance reads W + P. If the bird flies around in the box at a constant elevation without accelerating, what is the balance reading?

(a) P
(b) P - W
(c) P + 2W
(d) P + W/2
(e) P + W

$$F = P + \frac{W}{2} + \frac{W}{2} = P + W \; \bullet$$

Answer is (e)

STATICS 13

Given the truss as shown below. Determine mathematically the stress in members U_2U_3, U_2L_3, L_2L_3, and indicate if the member is in tension or compression.

Compute R_{L_0} and R_{L6}

$$\Sigma M_{L_0} = 0 \qquad (6 \times 15)R_{L6} - 15(4 \times 15) - 24(2 \times 15) = 0$$

$$R_{L6} = \frac{900 + 720}{90} = 18^k \qquad R_{L_0} = 24 + 15 - 18 = 21^k$$

$$\Sigma F_y = 0$$

$$F_2(\tfrac{4}{5}) + 21 - 24 = 0$$

$$F_2 = \frac{5(3)}{4} = 3.75^k$$

● $U_2L_3 = 3.75^k$ Comp.

$$\Sigma M_{U_2} = 0$$

$$20F_3 - 21(30) = 0$$

$$F_3 = 31.5^k$$

● $L_2L_3 = 31.5^k$ Tens.

$$\Sigma F_x = 0$$

$$F_1 + F_2(\tfrac{3}{5}) - F_3 = 0$$

$$F_1 = 31.5 - 3.75(\tfrac{3}{5})$$

$$F_1 = 31.5 - 2.25 = 29.25^k$$

● $U_2U_3 = 29.25^k$ Comp.

STATICS 14

A weight W_1 is supported by a rigid, weightless bar, AB, and a weightless cable loaded by a weight W_2, as shown. Find the relationship between W_1 and W_2 for equilibrium of the system, with the bar AB in a horizontal position, and show that this position of equilibrium is unstable. Neglect friction, and assume negligible dimensions for the pulley.

$$\Sigma F_y = 0$$

$$W_2 \cos 45° - W_1 = 0$$

$$W_1 = \frac{\sqrt{2}}{2} W_2$$

or $W_2 = \sqrt{2}\, W_1$ ●

In stable equilibrium small movements of the structure produce forces that tend to restore the structure to its original condition.

In unstable equilibrium small movements produce forces that tend to cause still further movement.

In this problem assume the structure is deflected by displacing B until bar AB makes a small angle ϕ with the horizontal. We can then compute the sum of the moments about joint A to see what the effect is on the structure.

<u>If B deflected downward</u>   <u>If B deflected upward</u>

clockwise moment = $W_1 L \cos \phi$ clockwise moment = $W_1 L \cos \phi$

counterclockwise moment: counterclockwise moment:

$$\sqrt{2}\, W_1 L \sin(45 - \frac{\phi}{2})$$ $$\sqrt{2}\, W_1 L \sin(45 + \frac{\phi}{2})$$

if ϕ small, say $5°$

$$\cos \phi = 0.996$$

$$\sin(45 - \frac{\phi}{2}) = 0.676$$

$$\sin(45 + \frac{\phi}{2}) = 0.737$$

$0.996 W_1 L > 0.956 W_1 L$ $0.996 W_1 L < 1.04 W_1 L$

CW moment $>$ CCW moment   CW moment $<$ CCW moment

   Unstable   Unstable

The structure is thus in unstable equilibrium. ●

STATICS 15

A piece of metal plate in the shape of a right triangle is suspended as shown below. If its weight is W, and its sides are a and b, what is the force at string A?

    (a) W/3
    (b) W/2
    (c) 2W/3
    (d) 3W/4
    (e) W

$$\Sigma M_B = 0 \qquad R_A(b) - W\left(\frac{2b}{3}\right) = 0$$

$$R_A = \frac{2bW}{3b} = \frac{2W}{3} \ ●$$

Answer is (c)

STATICS 16

Given a triangle as shown. Find the percentage error in the X coordinate of the centroid if five equal width rectangles are used to approximate the area. Assume line cd goes through the mid-point of each rectangle.

Compute the centroid of the 5 rectangles:

Rectangle	Height	Width	Area	Distance to Centroid	Ax
1	0.9h	0.2b	0.18bh	0.1b	$0.018b^2h$
2	0.7h	0.2b	0.14bh	0.3b	$0.042b^2h$
3	0.5h	0.2b	0.10bh	0.5b	$0.050b^2h$
4	0.3h	0.2b	0.06bh	0.7b	$0.042b^2h$
5	0.1h	0.2b	0.02bh	0.9b	$0.018b^2h$
			0.50bh		$0.170b^2h$

$$\bar{x} = \frac{\Sigma Ax}{\Sigma A} = \frac{0.170b^2h}{0.50bh} = 0.34b$$

We know, however, that the centroid of a triangle is located at $\frac{1}{3}b$

$$\text{Percent Error} = \frac{0.34b - 0.333b}{0.333b} = \frac{0.007b}{0.333b} = 2.1 \text{ percent} \bullet$$

STATICS 17

The reaction at B to hold the weightless hinged frame on the next page in equilibrium is:

(a) 333 lbs ⟶

(b) 333 lbs ⟶ plus 800 in. lbs. ⤸

(c) 333 lbs ⟶ plus 1200 in. lbs. ⤸

(d) 333 lbs   ╲ 36.9°

(e) 333 lbs   ╲ 53.1°

The line of action of all forces must pass through point O. Therefore the direction of R_B is

$$\alpha = \tan^{-1} \frac{4}{3} = 53.1^\circ \text{ with respect to the}$$

horizontal.

There must be a similar reaction at A.

The magnitude of reactions A and B:
$$R_A = R_B = \frac{533}{2} \times \frac{5}{4} = 333 \text{ lbs.}$$

Answer is (e) ●

STATICS 18

The moment of inertia of any plane figure can be expressed in units of length to the

    (a) First power
    (b) Second power
    (c) Third power
    (d) Fourth power
    (e) Fifth power

For a rectangular cross-section about its centroidal axis

$$I_x = \int y^2 \, dA = \int_{-h/2}^{+h/2} y^2 b \, dy = b \left[\frac{y^3}{3} \right]_{-h/2}^{+h/2} = \frac{bh^3}{12}$$

Thus the moment of inertia is a fourth power of length. ●

Answer is (d)

STATICS 19

For the beam loaded as shown, express the reactions at A and B by
algebraic expressions in terms of K and L.

$\Sigma M_A = 0$

$LB - \frac{1}{2}L(KL)(\frac{2}{3}L) = 0$ $B = \frac{KL^2}{3}$ ●

$\Sigma M_B = 0$

$-LA + \frac{1}{2}L(KL)(\frac{1}{3}L) = 0$ $A = \frac{KL^2}{6}$ ●

As a check

$\Sigma F_y = 0$ $A + B - \frac{1}{2}KL^2 = 0$ $\frac{1}{6}KL^2 + \frac{1}{3}KL^2 - \frac{1}{2}KL^2 = 0$ $0 = 0$

STATICS 20

The vector which represents the sum of a group of force vectors is called
the

   (a)   magnitude
   (b)   resultant
   (c)   sum
   (d)   phase angle
   (e)   force polygon

A vector sum is called the resultant vector. ●
                  Answer is (b)

STATICS 21

Which of the following is <u>not</u> a vector quantity?

   (a)   velocity
   (b)   speed
   (c)   acceleration
   (d)   displacement
   (e)   momentum

Speed is a rate $\frac{ds}{dt}$ and has a magnitude but not direction. It is not a
vector. ●
                  Answer is (b)

STATICS 22

A horizontal equilateral plate ABC (shown below), 10 feet on a side, is supported at the vertices. What are the reactions at the supports due to a load of 500 pounds acting at a point on the median line 3 feet from vertex A?

PLAN

500 LB.

ELEVATION

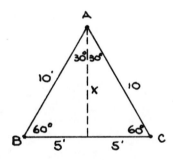

$$\cos 30° = \frac{X}{10} = 0.866$$

Therefore X = 8.66 ft

$\Sigma M_A = 0$

$$R_{B,C}(8.66) - 500(3) = 0 \qquad R_{B,C} = \frac{3}{8.66}(500) = 173 \text{ lbs}$$

$$R_B = R_C = \frac{173}{2} = 86.5 \text{ lbs} \;\bullet$$

$\Sigma F_y = 0$

$$R_{B,C} + R_A - 500 = 0 \qquad 173 + R_A - 500 = 0$$

$$R_A = 327 \text{ lbs} \;\bullet$$

STATICS 23

Find the centroid of a solid hemisphere of radius "r".

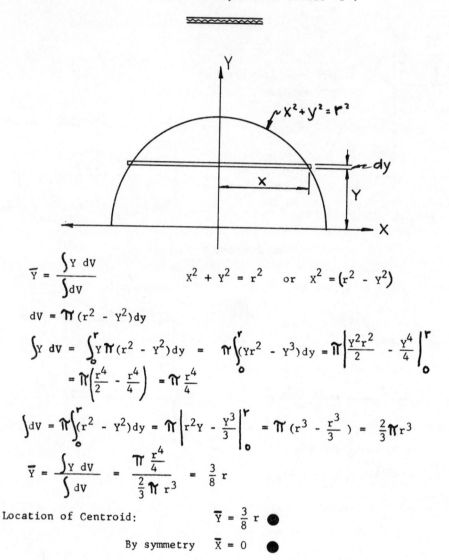

$$\overline{Y} = \frac{\int Y\,dV}{\int dV} \qquad\qquad x^2 + y^2 = r^2 \quad \text{or} \quad x^2 = \left(r^2 - y^2\right)$$

$$dV = \pi\left(r^2 - y^2\right)dy$$

$$\int Y\,dV = \int_0^r Y\,\pi\left(r^2 - y^2\right)dy = \pi\int_0^r \left(Yr^2 - y^3\right)dy = \pi\left|\frac{y^2 r^2}{2} - \frac{y^4}{4}\right|_0^r$$

$$= \pi\left(\frac{r^4}{2} - \frac{r^4}{4}\right) = \pi\,\frac{r^4}{4}$$

$$\int dV = \pi\int_0^r \left(r^2 - y^2\right)dy = \pi\left|r^2 Y - \frac{y^3}{3}\right|_0^r = \pi\left(r^3 - \frac{r^3}{3}\right) = \frac{2}{3}\pi r^3$$

$$\overline{Y} = \frac{\int Y\,dV}{\int dV} = \frac{\pi\,\frac{r^4}{4}}{\frac{2}{3}\pi r^3} = \frac{3}{8}\,r$$

Location of Centroid: $\qquad \overline{Y} = \frac{3}{8}\,r$ ●

$\qquad\qquad$ By symmetry $\quad \overline{X} = 0$ ●

3
DYNAMICS

DYNAMICS 1

If the disk shown below is rolling without slipping, what is the direction
of the velocity of point p relative to point q at the instant shown?

(a) ↑

(b) →

(c) ↓

(d) 45°

(e) 45°

Since p and q both lie on the disk, the facts that the disk contacts a
surface and that it rolls without slipping are unrelated to motion of p
relative to q.   p and q have equal tangential velocities, p directed
upward and q directed to the right.

For motion of p relative to q, consider q stationary and add q's reversed
velocity vector to p's vector. The resultant is motion of p relative
to (fixed) q.

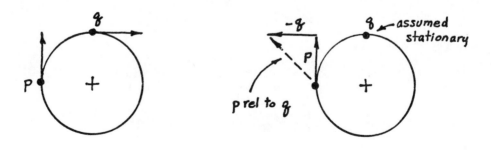

Answer is (e) ●

DYNAMICS 2

A 3.22-pound bead slides down a wire without friction along a circular path in a vertical plane as shown below. The speed of the bead along the wire in the position shown is 10.0 feet per second. The magnitude of the force of the wire on the bead in this position is:

(a)   0.576 lb.

(b)   1.93   lb.

(c)   2.00   lb.

(d)   2.57   lb.

(e)   3.22   lb.

Since there is no friction between bead and wire, there is no tangential component of bead weight exerted on the wire. Total force on wire is the vector sum of the radial component of bead weight and the radial force caused by its tangential velocity of 10 ft/sec.

$$F_{radial} = \frac{mv_T^2}{r} = \frac{3.22(10)^2}{32.2(5)} = 2.00 \text{ lb} \nearrow$$

Radial and tangential components of bead weight are:

$$\text{tangential} = \frac{3}{5}(3.22) = 1.93 \text{ lb} \searrow$$

$$\text{radial} = \frac{4}{5}(3.22) = 2.58 \text{ lb} \nearrow$$

W = 3.22 lb

Net force of bead on wire = 2.58 lb $\nearrow$ + 2.00 lb $\nearrow$ = 0.58 lb $\nearrow$

Reaction force of wire on bead = 0.58 lb $\nearrow$ ●

Answer is (a)

DYNAMICS 3

A large coil spring (k = 10 lb/in) is elongated, within its elastic range, by one foot. Stored energy in foot-pounds is nearest to:

(a)   10

(b)   40

(c)   60

(d)   80

(e)   120

F = kx,   k = 10 lb/in = 120 lb/ft.

Energy, $E = \int_0^1 F dx = \int_0^1 kx\,dx = 120 \left.\frac{x^2}{2}\right]_0^1 = 60 \text{ ft lb}$ ●

Answer is (c)

DYNAMICS 4

A ball thrown from A with velocity 17 ft/sec strikes a fixed barrier at B, and rebounds to C at a velocity of 16.25 ft/sec. ϕ is unequal to ϕ'

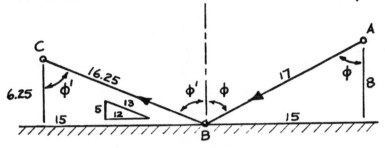

The coefficient of restitution is nearest to:

(a) 0.74
(b) 0.78
(c) 0.83
(d) 0.92
(e) 0.96

In this particular situation, coefficient of restitution $e = \dfrac{\tan \phi}{\tan \phi'}$

where ϕ is the incident angle and ϕ' is the emergent angle.

$$e = \frac{15/8}{12/5} = \frac{25}{32} = 0.78$$

In the usual case where both impacting objects move:

$$e = \frac{\text{relative velocity after impact}}{\text{relative velocity before impact}} = \frac{v_2' - v_1'}{v_1 - v_2}$$

For oblique impact the velocities used must be the components of velocity in a direction normal to the point of contact, in this case, the y direction.

$$e = \frac{0 - (+6.25)}{-8 - 0} = 0.78 \; \bullet$$

Answer is (b)

DYNAMICS 5

The cords and frictionless pulley, of the system shown in the drawing, have negligible mass. The distance in feet that the 32.2 lb mass will move during the first half second after release from rest is nearest to:

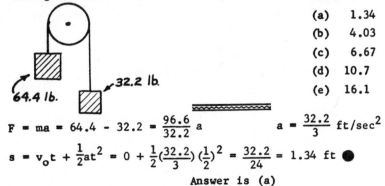

(a) 1.34
(b) 4.03
(c) 6.67
(d) 10.7
(e) 16.1

$$F = ma = 64.4 - 32.2 = \frac{96.6}{32.2}\, a \qquad a = \frac{32.2}{3} \text{ ft/sec}^2$$

$$s = v_o t + \frac{1}{2}at^2 = 0 + \frac{1}{2}\left(\frac{32.2}{3}\right)\left(\frac{1}{2}\right)^2 = \frac{32.2}{24} = 1.34 \text{ ft} \; \bullet$$

Answer is (a)

DYNAMICS 6

Given the velocity-time plot for motion shown below:

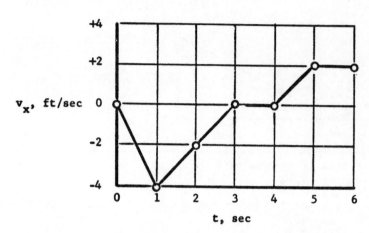

v_x, ft/sec

t, sec

Displacement of the object from t = 1 sec to t = 6 sec will be:

        (a)   4 feet to left
        (b)   1 foot to left
        (c)   zero
        (d)   1 foot to right
        (e)   3 feet to right

$S_x = \int_1^6 v_x dt.$ Graphically integrate by summing positive and negative areas, above and below the zero velocity line, between t = 1 and t = 6.

$$S_x = -4 + 1 + 2 = -1 \text{ (to left)} \ \bullet$$

Total travel will have been 7 feet (4 to left and 3 to right).

Answer is (b)

DYNAMICS 7

A hoist with a 100-horsepower engine is capable of lifting a 10,000 lb load a height of 20 feet in 10 seconds. What is the efficiency of this machine?

$$\text{output horsepower} = \frac{10,000 \text{ lbs} \times \dfrac{20 \text{ feet}}{10 \text{ sec.}}}{550 \ \dfrac{\text{ft-lb}}{\text{sec.}}} = 36.4 \text{ hp}$$

$$\text{efficiency} = \frac{\text{output hp}}{\text{input hp}} = \frac{36.4}{100} = 0.364 = 36.4 \text{ per cent} \ \bullet$$

DYNAMICS 8

An object is placed 3 feet from the center of a horizontally rotating platform. The coefficient of friction is 0.30. The object will begin to slide off when the platform speed is nearest to:

       (a)   12 rpm
       (b)   17 rpm
       (c)   22 rpm
       (d)   26 rpm
       (e)   33 rpm

$a_r = r\omega^2 = 3\omega^2$, where ω is rad/sec.

W = weight of object

$$F_r - F_F = 0 = \frac{W}{32.2}\, 3\omega^2 - 0.3W$$

$$0 = \frac{3}{32.2}\,\omega^2 - 0.3$$

$$\frac{3\,\omega^2}{32.2} = 0.3 \qquad \therefore \omega^2 = \frac{0.3(32.2)}{3}$$

$$\omega = \sqrt{3.22} = 1.79 \text{ rad/sec}$$

$F_r = ma_r$

$F_F = 0.3W$

$$1.79\,\frac{\text{rad}}{\text{sec}} \times \frac{60\text{ sec}}{\text{min}} \times \frac{\text{rev}}{2\pi\text{ rad}} = 17.1\,\frac{\text{rev}}{\text{min}} \; \bullet$$

W

Answer is (b)

DYNAMICS 9

Two barges, one weighing 10 tons, the other weighing 20 tons are connected by a cable in quiet water. Initially the barges are at rest 100 feet apart. The cable is reeled in until the barges are 50 feet apart. If the friction is negligible, calculate the distance moved by the 10 ton barge.

Equal and opposite forces are exerted on A and B when the cable is reeled in. Assuming constant force, accelerations are constant, but unequal.

$$F_A = F_B = m_A a_A = m_B a_B \qquad \frac{20,000}{32.2}\,a_A = \frac{40,000}{32.2}\,a_B \qquad \therefore a_A = 2a_B$$

At any time t, the distances traveled from rest are:

$$s_A = \frac{1}{2}\,a_A t^2 \quad \text{and} \quad s_B = \frac{1}{2}\,a_B t^2$$

Therefore

$$\frac{s_A}{s_B} = \frac{a_A}{a_B} = \frac{2a_B}{a_B} = \frac{2}{1} \qquad \text{Since } s_A + s_B = 50 \text{ ft}, \qquad s_A = \frac{2}{3}(50) = 33.3 \text{ ft} \; \bullet$$

DYNAMICS 10

A 60-ton car is to travel around a curve of radius 3,000 ft. at a speed of 60 mph. By how much must the outer rail be elevated in order that the reaction against the track may be perpendicular to the plane of the rails? The track is standard gage of 4'-8 1/2".

Car weight W is directed vertically downward through its cg. Reaction against track is the resultant of radial force and car weight. Resultant must be perpendicular to the plane of the rails.

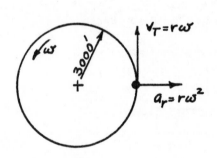

Tangential velocity, v_T, is 60 mph = 88 ft/sec.

Radial acceleration $a_r = r\omega^2$

Radial Force $F_r = ma_r = mr\omega^2 = \dfrac{mv_T^2}{r}$

$$F_r = \frac{W\,(88)^2}{32.2\,(3000)} = \frac{7,750W}{96,600}$$

$$= 0.08W$$

$\tan\theta = \dfrac{0.08W}{W} = 0.08$ (For small angles $\sin\theta = \tan\theta$)

Therefore,

Elevation = 56.5(0.08) = 4.53" ●

DYNAMICS 11

Dimensions of the reciprocating engine in the drawing are crank equals 2 inches and connecting rod equals 6 inches.

The crank rotates at a constant angular velocity of 10 rad/sec.

Find the instantaneous velocity and acceleration of the piston at the position shown.

This is a problem involving combined rotation, and rotation plus translation.

Since the journal bearing at A connects crank OA and connecting rod AB, motion of A derived from its constant velocity rotation about O is the same as motion obtained from translation of piston B plus rotation of the connecting rod AB about wrist pin B.

Solution depends upon equating x and y components that result both from translation of B and from A's rotation about B.

Angle ϕ is determined by law of sines for triangle OAB:

$$\frac{\sin \phi}{2"} = \frac{\sin 30°}{6"} \qquad \sin \phi = \frac{2}{6}(0.50) = \frac{1}{6} = 0.167, \quad \text{and from a right}$$

triangle of sides 1, 6 and $\sqrt{35}$: $\quad \cos \phi = \frac{\sqrt{35}}{6} = 0.987$

Curvilinear motion involves normal N (radial) and tangential T components of velocity and acceleration:

$$v_T = r \omega \qquad\qquad a_T = r \alpha$$
$$(\alpha = 0 \text{ at constant } \omega)$$

$$v_N = \frac{dr}{dt} = 0, \text{ at constant radius}$$

$$a_N = r \omega^2, \text{ directed toward center of rotation}$$

These components are shown on the vector diagram, along with x and y coordinate axes.

Primes (') designate items involved in rotation about B, while unprimed items involve rotation about O.

Directions of tangential velocity and acceleration vectors must be consistent with stated or assumed directions of α and ω producing them.

Positive answers indicate correct direction assumption; negative answers indicate the opposite direction.

The x component of velocity of A, v_{Ax}, results from sum of x components of v_T and v_N. v_{Ax} also equals sum of x components of v_T' and v_N' plus x component of B's translation, v_{Bx}. Since radii are fixed and B is constrained to only translate vertically, v_N, v_N', and v_{Bx} are zero.

$$v_{Ax} = v_T \cos 30° - v_N \sin 30° = v_T' \cos \phi - v_N' \sin \phi + v_{Bx}$$

$$r\,\omega \cos 30° - 0 = r'\,\omega' \cos \phi - 0 + 0$$

$$2(10)(0.866) = 6(\omega')(0.987)$$

$$\omega' = + 2.93 \text{ rad/sec} \;\;\; \text{(angular velocity at B)}$$

Similarly, y-component of A's velocity equals sum of y components of v_T and v_N, and also equals sum of y components of v_T' and v_N' plus y component of B's translation.

$$v_{Ay} - v_T \sin 30° - v_N \cos 30° = v_T' \sin \phi + v_N' \cos \phi - v_{By}$$

$$- r\,\omega \sin 30° - 0 = r'\,\omega' \sin \phi + 0 - v_{By}$$

$$-2(10)(0.500) = 6(2.93)(0.167) - v_{By}$$

$$v_{By} = + 12.93 \text{ in/sec} \;\;\; \bullet \text{ (piston velocity)}$$

Accelerations are similarly determined from x and y components at A. α is zero, but α' is not.

x component: $a_T \cos 30° - a_N \sin 30° = a_T' \cos \phi - a_N \sin \phi + a_{Bx}$

$$r \alpha \cos 30° - r \omega^2 \sin 30° = r' \alpha' \cos \phi - r' (\omega')^2 \sin \phi + 0$$

$$2(0)(0.866) - 2(10)^2(0.500) = 6(\alpha')(0.987) - 6(2.93)^2(0.167) + 0$$

$$\alpha' = -15.45 \text{ rad/sec}^2 \;\curvearrowright$$

(angular acceleration at B)

y component: $-a_T \sin 30° - a_N \cos 30° = a_T' \sin \phi + a_N' \cos \phi - a_{By}$

$$0 \sin 30° - r \omega^2 \cos 30° = r' \alpha' \sin \phi + r' (\omega')^2 \cos \phi - a_{By}$$

$$0 - 2(10)^2(0.866) = 6(-15.45)(0.167) + 6(2.93)^2(0.987) - a_{By}$$

$$a_{By} = +208.5 \text{ in/sec}^2 \downarrow \;\bullet$$

(piston acceleration)

DYNAMICS 12

A system of four particles of equal mass (m) rotate with an angular velocity (ω) at equal distances (L) from the center of rotation, and they are spaced at equal angles as shown.
Find the magnitude and direction of the vector, $\vec{H}_c$, representing the moment of momentum of the system about the point of rotation.

$\vec{H}_c$ = constant vector

$$= 4(\vec{\rho} \times m \vec{V}) = 4m\rho V \sin 90° = 4 m \rho (\rho \omega)$$

Since $\rho = L$ $\therefore \vec{H}_c = 4mL^2\omega \;\bullet$

$\vec{H}_c$ is perpendicular to the plane of ρ and V, and is directed outward from the figure, i.e., toward the reader. $\bullet$

DYNAMICS 13

A 10-ton space craft orbits the earth at a mean altitude of 200 miles with a mean tangential velocity of 17,000 mph. At what velocity will the craft orbit the moon at the same elevation above the surface?

Diameter of the earth - - - - 8,000 miles

Diameter of the moon   - - - - 2,000 miles

Mass of the earth - - - - - - 5.4 x 10²¹ tons

Mass of the moon - - - - - - 1/81 that of the earth

 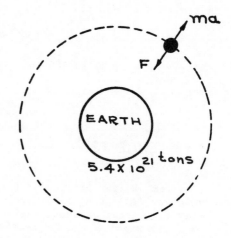

$$F = G \frac{M m}{R^2}$$ where F = gravitational attraction

G = gravitational constant

M and m = masses

R = distance between centers of the two masses

For equilibrium the gravitational attraction must be counterbalanced by the centrifugal force.

$$G \frac{M m}{R^2} = mass(a) = m \frac{v^2}{R}$$

Solving for v^2 $v^2 = \frac{GM}{R}$

Earth Orbit   Moon Orbit

$$V_e^{\ 2} = \frac{GM_e}{R_e} \quad (1)$$ $$V_m^{\ 2} = \frac{GM_m}{R_m} \quad (2)$$

Dividing $\frac{(2)}{(1)}$

$$\frac{V_m^{\ 2}}{V_e^{\ 2}} = \frac{\frac{GM_m}{R_m}}{\frac{GM_e}{R_e}} = \frac{R_e M_m}{R_m M_e}$$

$$V_m = \sqrt{\frac{V_e^{\ 2} R_e M_m}{R_m M_e}} = \sqrt{\frac{17,000^2 (4000 + 200)(1)}{(1000 + 200)(81)}}$$

$$= \sqrt{12.5 \times 10^6} = 3,530 \text{ mph} \ \bullet$$

DYNAMICS 14

A projectile weighing 100 pounds strikes the concrete wall of a fort with a impact velocity of 1200 feet per second. The projectile comes to rest in 0.01 second, having penetrated the 8-foot thick wall to a distance of 6 feet. What is the average force exerted on the wall by the projectile?

Knowing that the impulse exerted on the wall is equal to the change of momentum of the projectile, one may write

$$\int F \, dt = m(V_1 - V_2)$$

Assuming F is the average force exerted, it can be considered as a constant, thus the equation becomes:

$$Ft = m(V_1 - V_2) \qquad F(0.01) = \frac{100}{32.2}(1200 - 0)$$

$$F = 3.73 \times 10^5 \text{ pounds} \quad \bullet$$

DYNAMICS 15

A 10" diameter pulley is belt driven with a net torque of 250 ft-lbs. The ratio of tensions in the tight to slack sides of the belt is 4 to 1. What is the maximum tension in the belt?

(a) 250 lbs
(b)   83 lbs
(c) 800 lbs
(d) 500 lbs
(e) 333 lbs

$$\frac{5}{12} T - \frac{5}{12} \cdot \frac{T}{4} = 250 \text{ ft-lbs}$$

$$\frac{3}{4} \cdot \frac{5T}{12} = 250$$

$$T = \frac{250 \times 12 \times 4}{3 \times 5} = 800 \text{ lbs.} \bullet$$

Answer is (c)

DYNAMICS 16

Two 3-lb weights are connected by a massless string hanging over a smooth, frictionless peg. If a third weight of 3 lbs is added to one of the weights and the system is released, by how much is the force on the peg increased?

Initial condition:

T = 3 lbs.

2T = 6 lbs.

After additional weight added:

$$\Sigma F = ma$$

$$6 - T_1 = \frac{6}{32.2} a \quad (1)$$

$$T_1 - 3 = \frac{3}{32.2} a \quad (2)$$

Solving these two equations

from (2)
$$a = \frac{32.2}{3}(T_1 - 3)$$

substituting into (1)

$$6 - T_1 = \frac{6}{32.2} \frac{32.2}{3}(T_1 - 3)$$

$$6 - T_1 = 2(T_1 - 3) \qquad 3T_1 = 12 \qquad T_1 = 4 \text{ lbs}$$

The force on the peg is increased by the quantity $2T_1 - 2T$

$$= 8 - 6 = 2 \text{ lbs.} \quad \bullet$$

DYNAMICS 17

A body experiences acceleration "a" given by the expression
$a = At - Bt^2$, where A and B are constants and t is time.
If, at time $t = 0$, the body has zero displacement and velocity,
at what next value of time does the body again have zero displacement?

$$a = At - Bt^2 \qquad \frac{dv}{dt} = a \qquad \frac{ds}{dt} = v$$

The differential equation is

$$\frac{dv}{dt} = At - Bt^2 \qquad \text{boundary conditions} \quad t=0, \; v=0.$$

Solving
$$v = A\frac{t^2}{2} - B\frac{t^3}{3} + C \qquad \text{where C is a constant of integration}$$

Using the boundary conditions, $\quad v = A(0) - B(0) + C = 0 \qquad \therefore C = 0$

$$v = A\frac{t^2}{2} - B\frac{t^3}{3}$$

The new differential equation is

$$\frac{ds}{dt} = A\frac{t^2}{2} - B\frac{t^3}{3} \qquad \text{boundary conditions} \quad t=0, \; s=0.$$

Solving
$$s = A\frac{t^3}{6} - B\frac{t^4}{12} + D \qquad \text{where D is a constant of integration}$$

Using boundary conditions, $\quad s = 0 = A(0) - B(0) + D \qquad \therefore D = 0$

Therefore
$$s = A\frac{t^3}{6} - B\frac{t^4}{12}$$

$$s = \left(\frac{A}{6} - \frac{Bt}{12}\right)t^3$$

Setting s = 0

$$s = (\frac{A}{6} - \frac{Bt}{12})t^3 = 0$$

$$\frac{A}{6} = \frac{Bt}{12} \qquad t = \frac{2A}{B} \quad \bullet$$

DYNAMICS 18

A balloon is rising vertically at the rate of 55 feet per second. When the balloon is 1000 feet above the surface of the earth (flat surface) a bullet is fired from it downward at a 30 degree angle with the vertical. The bullet has a muzzle velocity of 1100 feet per second.

Compute the vertical component of the velocity of the bullet as it strikes the earth.

Note: $g = 32.2$ ft/sec^2. Neglect air friction.

Let the downward direction be positive

$$y = 1000 \text{ feet} \qquad a_y = 32.2 \text{ ft/sec}^2$$

$$\frac{d v_y}{dt} = a_y \qquad \therefore v_y = \int a_y dt = a_y t + C$$

imposing the initial conditions

$$v_y(0) = 1100 \cos 30^\circ - 55$$

$$\therefore C = 953 - 55 = 898$$

$$v_y(t) = 898 + a_y t$$

$$y(t) = \int v_y dt = 898t + \frac{a_y t^2}{2} + C_2$$

Measuring from the balloon $y(0) = 0 \therefore C_2 = 0$

At impact with the earth $y = 1000' = 898t + \frac{32.2t^2}{2}$

or $16.1t^2 + 898t - 1000 = 0$

Solving the quadratic

$$t = \frac{-898 \pm \sqrt{(898)^2 - 4(16.6)(-1000)}}{32.2}$$

$$= \frac{-898 \pm \sqrt{87 \times 10^4}}{32.2} = -27.9 \pm 29.0$$

$t = +1.1$ seconds at impact (the negative root has no physical meaning).

At impact

$$v_y(t) = 898 + 32.2(1.1)$$

$$v_y(1.1) = 933. \text{ feet/second} \quad \bullet$$

DYNAMICS 19

If the radius of the earth is approximately 4 times the radius of the moon, and if the mass of the moon is 1/82 of that of the earth, calculate the weight of a pound-mass on the surface of the moon.

The force of attraction between two masses may be computed from the equation:

$$F = G \frac{Mm}{d^2}$$ where G is the gravitational constant

M and m are the masses of the two bodies

d is the distance between the two bodies

1 lbm on the surface of the earth has a weight of 1 lbf, therefore

$$1 = G \frac{M_{earth}(1 \text{ lbm})}{(4R)^2} \qquad (1)$$

Applying this equation now to 1 lbm on the surface of the moon gives:

$$F_{moon} = G \frac{\frac{1}{82}M_{earth}(1 \text{ lbm})}{R^2} \qquad (2)$$

dividing (2) by (1)

$$F_{moon} = \frac{\dfrac{GM_{earth}(1 \text{ lbm})}{82R^2}}{\dfrac{GM_{earth}(1 \text{ lbm})}{16R^2}} = \frac{16}{82} = 0.195 \text{ lbf} \; \bullet$$

DYNAMICS 20

A flywheel weighing 4,480 lbs has a radius of gyration of 3 feet. If there is a driving torque of 600 lb-ft and a resisting torque of 250 lb-ft, find the time required to increase its speed from 20 to 80 revolutions per minute.

Change in angular velocity = $(80 - 20)\dfrac{2\pi}{60}$ = 6.28 radians/second = $\Delta\omega$

Moment of Inertia (I) = mr^2 where m = mass

r = radius of gyration (sometimes designated k)

$$I = mr^2 = \frac{W}{g}r^2 = \frac{4480}{32.2}(3)^2 = 1252 \text{ ft-lb-sec}^2$$

Solving by Impulse-Momentum for Rotation:

$$\int_o^t \Sigma M \cdot dt = I\omega - I\omega_o$$

For ΣM = constant, the equation becomes

$$\Sigma Mt = I(\omega - \omega_o)$$

Substituting the values given for this problem

$$(600 - 250)t = 1252(6.28) \qquad t = \frac{1252(6.28)}{350} = 22.5 \text{ seconds} \; \bullet$$

DYNAMICS 21

A pendulum with a concentrated mass (m) is suspended vertically inside a stationary railroad freight car by means of a rigid weightless connecting rod. If the connecting rod is pivoted where it attaches to the box car, compute the angle that the rod makes with the vertical as a result of a constant horizontal acceleration of 2 feet/sec^2 of the box car.

$$\tan \alpha = \frac{ma}{mg} = \frac{a}{g} = \frac{2}{32.2} = 0.0621$$

$$\alpha = 3.55° \ \bullet$$

DYNAMICS 22

A body moves so that during the first part of its motion its distance S in inches from the starting point is given by the expression

$$S = 6.8t^3 - 10.8t \quad (t \text{ in seconds})$$

The acceleration in 3 seconds would be

    (a) 172.8 inches/sec^2
    (b) 122.4 "   "
    (c)   61.2 "   "
    (d) 212.4 "   "
    (e) none of the above

$$S = 6.8t^3 - 10.8t$$

$$\frac{ds}{dt} = 3(6.8)t^2 - 10.8 = 20.4t^2 - 10.8$$

$$\frac{d^2s}{dt^2} = 2(20.4)t = 40.8t$$

At t = 3 seconds

$$\frac{d^2s}{dt^2} = 40.8(3) = 122.4 \text{ inches/sec}^2 \ \bullet$$

Answer is (b)

DYNAMICS 23

A heavy brass plumb-bob suspended from a 38-inch cord was observed to have a natural period of oscillation of about two seconds if pulled 24 inches to one side and then allowed to swing freely. If an astronaut was to repeat this experiment on the moon where the gravitational attraction is approximately 1/6th that of the earth, the observed period would be:

    (a) 0.33 seconds
    (b) 0.82 "
    (c) 4.90 "
    (d) 12.00 "
    (e) 72.00 "

$$T = 2\pi\sqrt{\frac{L}{g}}$$

where L = length of pendulum (radius of the arc)
         g = acceleration of gravity

The radius of the arc L probably is greater than 38 inches.
Solving for L

$$2 = 2\pi\sqrt{\frac{L}{32.2}} \qquad L = \frac{32.2}{\pi^2} = 3.26 \text{ feet} \quad (39.1 \text{ inches})$$

On the moon

$$g_{moon} = \frac{g}{6}$$

$$T = 2\pi\sqrt{\frac{3.26}{\frac{g}{6}}} = 2\pi\sqrt{\frac{6 \times 3.26}{32.2}} = 4.90 \text{ seconds} \bullet$$

Answer is (c)

DYNAMICS 24

A projectile launched at 2000 ft/sec from A impacts at the same elevation at B. Assume no in-flight propulsion and no aerodynamic drag.
Horizontal distance in miles between A and B is nearest to:

    (a)   7.9
    (b) 10.2
    (c) 15.9
    (d) 20.3
    (e) 24.9

Absence of drag allows horizontal component of velocity to remain at 2000 cos 30° = + 1732 ft/sec throughout flight. Vertical component, initially at 2000 sin 30° = + 1000 ft/sec, is decelerated by gravity to zero vertical velocity at maximum altitude, then accelerated to -1000 ft/sec down at impact. Time of flight is twice the time to reach maximum altitude.

$$v_y = v_{oy} - gt = +1000 - 32.2t = 0 \qquad t = 31 \text{ sec} \qquad 2t = 62 \text{ sec}$$

$$s_x = v_x(2t) = 1732 \times 62.0 = 107,000 \text{ ft} = \frac{107,000}{5,280} = 20.3 \text{ miles} \bullet$$

Answer is (d)

4
MECHANICS
OF MATERIALS

MECH OF MTLS 1

Two circular shafts, one hollow and one solid, are made of the same material and have the diameters shown below. If T_h is the twisting moment that the hollow shaft can resist and T_s is the twisting moment that the solid shaft can resist, the ratio of T_h to T_s is:

(a) 1/4

(b) 1/2

(c) 9/16

(d) 15/16

(e) 1

Knowing the torsion formula $T = \dfrac{2S_s J}{D}$

Compare the polar moments of inertia, J

for circular cross section $J = \dfrac{\pi D^4}{32}$

for hollow cross section $J = \dfrac{\pi(D^4 - d^4)}{32} = \dfrac{\pi}{32}(D^4 - \dfrac{D^4}{16}) = \dfrac{\pi D^4}{32}(\dfrac{15}{16})$

Establish ratio:

$$\frac{T_h}{T_s} = \frac{\dfrac{2}{D} S_s \dfrac{\pi D^4}{32} \dfrac{15}{16}}{\dfrac{2}{D} S_s \dfrac{\pi D^4}{32}} = \frac{15}{16} \quad \bullet$$

Answer is (d)

MECH OF MTLS 2

A steel cylinder 12 inches outside diameter and a wall thickness of 1 inch is filled with concrete and used as a pier to support an axial load in compression. If the allowable stresses are 40,000 psi for steel and 3,500 psi for concrete and the Youngs modulus for the steel is 30×10^6 psi and for the concrete is 2.8×10^6 psi, what is the allowable load on the pier?

(Assume both materials have linear relationship between stress and strain.)

Calculate cross sectional areas

   concrete: $\frac{\pi}{4} d^2 = \frac{100\pi}{4} = 25\pi$ in^2

   steel: $\frac{\pi}{4}(144) - 25\pi = 11\pi$ in^2

In a composite structure maximum allowable strain ϵ of the composite is limited to the lesser of the individual allowable strains.

   concrete: $\epsilon_{max} = \frac{S_c}{E} = \frac{3.5 \times 10^3}{2.8 \times 10^6} = 1.25 \times 10^{-3}$ in/in

   steel: $\epsilon_{max} = \frac{S_c}{E} = \frac{40 \times 10^3}{30 \times 10^6} = 1.33 \times 10^{-3}$ in/in

Both materials are therefore limited to $\epsilon = 1.25 \times 10^{-3}$ in/in and this gives us a reduced allowable steel compressive stress

   $S_c = \frac{1.25 \times 10^{-3}}{1.33 \times 10^{-3}} \times 40,000 = 37,500$ psi

Total allowable load is the summation of steel and concrete loads (to the allowable ϵ).

Total P = 3,500(25π) + 37,500(11π) = 499,500 π = 1,570,000 lb ●

MECH OF MTLS 3

The hollow steel shaft 100 inches long   must transmit a torque of 300,000 in-lb. The total angle of twist must not exceed 3 degrees. The maximum shearing stress must not exceed 16,000 psi. Find the Inside Diameter (d) and Outside Diameter (D) of the shaft that meets these conditions.

Shear modulus for steel
   $G = 12 \times 10^6$ psi.

Use torsion formula $T = \dfrac{2 S_s J}{D}$

Polar Moment of Inertia:

solid shaft: $J = \dfrac{\pi}{32} D^4$

hollow shaft: $J = \dfrac{\pi}{32} (D^4 - d^4)$

Shear modulus $G = \dfrac{S_s}{\phi} = \dfrac{\text{shear stress,}}{\text{shear strain,}} \dfrac{\text{psi}}{\textbf{rad.}}$

For small angles:

θ in radians $= \sin \theta$

$\theta = 3^\circ = \dfrac{3}{57.3}$

$= 0.0524$ rad.

$\phi = \dfrac{r\theta}{L} = (\dfrac{D}{2})\dfrac{0.0524}{100} = 0.000262 \, D$

Actual shear stress (based on angle of twist):

$S_s = G \phi = 12 \times 10^6 \times 262 \times 10^{-6} \, D = 3,144 D$ psi

Allowable shear stress = 16,000 psi

$D = \dfrac{16,000}{3,144} = 5.10"$ ● This is twist limited; it may or may not carry the required torque.

Determine that this diameter of <u>solid shaft</u> will transmit the required torque of 300,000 in-lb.

$T = \dfrac{2(16,000)}{D} \dfrac{\pi}{32} D^4 = 1000 \pi D^3 = 1000 \pi (5.10)^3 = 418,000$ in-lb.

This is more than adequate, therefore the shaft may be hollowed out for weight reduction.

Use the hollow shaft polar moment of inertia formula in the torsion formula with D = 5.10" and solve for d.

$300,000 = \dfrac{2(16,000)}{D} \dfrac{\pi}{32}(5.10^4 - d^4)$

$\dfrac{300,000(5.10)}{1000 \pi} = 676 - d^4, \qquad d^4 = 188, \qquad d = 3.70"$ ●

MECH OF MTLS 4

A distance from A to B was measured to be 5,368.25' with a steel tape. The temperature during this time was 22°F. The tape was a standard 100.00' at 68°F. If the coefficient of expansion is 0.0000065, what is the true distance between A and B?

$\Delta L = \propto L \, \Delta T$ where $\propto$ = coefficient of expansion
L = 100 feet
$\Delta T = (22 - 68) = -46^{\circ}$

$\Delta L = 6.5 \times 10^{-6} \times 100(22 - 68) = -0.0298$ feet

Thus at $22^{\circ}F$ the tape was 0.0298 feet shorter than 100 feet.

The tape was used $\dfrac{5368.25}{100} = 53.6825$ times to measure the distance.

$53.6825 \times 0.0298 = 1.6$ feet

Therefore the distance A to B = 5368.25 - 1.6
= 5366.65 feet ●

MECH OF MTLS 5

The stress in an elastic material is

    (a)   inversely proportional to the material's yield strength
    (b)   inversely proportional to the force acting
    (c)   proportional to the displacement of the material acted upon by the force
    (d)   inversely proportional to the strain
    (e)   proportional to the length of the material subject to the force

According to Hooke's law: Stress is directly proportional to strain. Strain is the deformation (or displacement) of the material per unit of length. Thus we can say: Stress is proportional to the displacement of the material acted upon by the force. ●

Answer is (c)

MECH OF MTLS 6

The "SLENDERNESS RATIO" of a column is generally defined as the ratio of its

    (a)   length to its minimum width
    (b)   unsupported length to its maximum radius of gyration
    (c)   length to its moment of inertia
    (d)   unsupported length to its least radius of gyration
    (e)   unsupported length to its minimum cross-sectional area

In Euler's equation

$$\sigma_{crit} = \frac{\pi^2 E}{(L/r)^2}$$
where   L = unsupported length
r = least radius of gyration

the ratio L/r is called the column <u>slenderness</u> <u>ratio</u>. ●

Answer is (d)

MECH OF MTLS 7

The relationship between the extension of a spring and the force required to produce the extension is

$$\text{(a)} \quad F = ma$$
$$\text{(b)} \quad F = \mu N$$
$$\text{(c)} \quad F = \frac{mv^2}{R}$$
$$\text{(d)} \quad F = kx$$
$$\text{(e)} \quad F = \mu H$$

Hooke's law, that is applicable to springs, is:

$$F = kx \quad \text{where } F = \text{force}$$
$$k = \text{spring constant}$$
$$x = \text{change in length}$$

Answer is (d) ●

MECH OF MTLS 8

The linear portion of the stress-strain diagram of steel is known as the

(a) modulus of elongation
(b) plastic range
(c) irreversible range
(d) elastic range
(e) range of yield points

The portion of the diagram where stress is linearly proportional to strain is called the elastic range. ●

Answer is (d)

MECH OF MTLS 9

A spring has a natural length of 12" when at rest. If a force of 10 lbs is required to stretch it 1", find the work required (in ft-lb) to stretch the spring from a total length of 13" to a length of 15".

Using Hooke's law: $F = kx$ where F = force
k = spring constant
x = change of length

If a force of 10 lbs will cause a 1" change of length then

$$k = \frac{F}{x} = \frac{10 \text{ lbs}}{1 \text{ inch}} = 10 \text{ lbs/inch}$$

$$W = \int F \, ds = \int_{1}^{3} 10s \, ds = \left. \frac{10s^2}{2} \right|_{1}^{3} = 5(9 - 1) = 40 \text{ inch-lbs}$$

$$= \frac{40 \text{ inch-lbs}}{12 \text{ inchs/ft}} = 3 \frac{1}{3} \text{ ft-lbs.} \ ●$$

MECH OF MTLS 10

The following compression strengths were recorded for five test cylinders
taken from one batch of concrete:

> (a) 5,100 psi
> (b) 5,620 psi
> (c) 5,290 psi
> (d) 5,380 psi
> (e) 6,110 psi

REQUIRED: Determine the coefficient of variation of these results when
the standard deviation, σ, is defined as

$$\sigma = \sqrt{\frac{\sum (x - \bar{x})^2}{n-1}}$$

Standard deviation, σ, is a measure of the dispersion or scatter of a
set of values. It is sometimes called the root-mean-square deviation
for this is its method of calculation. First, square the deviations of
individual values from the arithmetic mean. Then take the mean of these
squares and extract the square root.

$$
\begin{array}{l}
5,100 \\
5,620 \\
5,290 \\
5,380 \\
6,110 \\
\end{array}
$$

$\sum x = 27,500$ arithmetic mean $= \dfrac{27,500}{5} = 5,500 = \bar{x}$

	$x - \bar{x}$	$(x - \bar{x})^2$
$5,100 - 5,500 =$	-400	$160,000$
$5,620 - 5,500 =$	120	$14,400$
$5,290 - 5,500 =$	-210	$44,100$
$5,380 - 5,500 =$	-120	$14,400$
$6,110 - 5,500 =$	610	$372,100$
$\sum x = 27,500$	0	$605,000 = \sum (x - \bar{x})^2$

Standard Deviation

$$\sigma = \sqrt{\frac{605,000}{4}} = 389$$

Coef. of Variation $= \dfrac{\text{Std Dev}}{\text{mean}} \times 10^2 = \dfrac{389}{5,500} \times 10^2 = 7.07\%$ ●

MECH OF MTLS 11

Shown on the next page is a frame with the loads and constraints as
indicated on the diagram.

REQUIRED: By means of a graphical solution, determine magnitude and
direction of the reactions at R_1 and R_2.

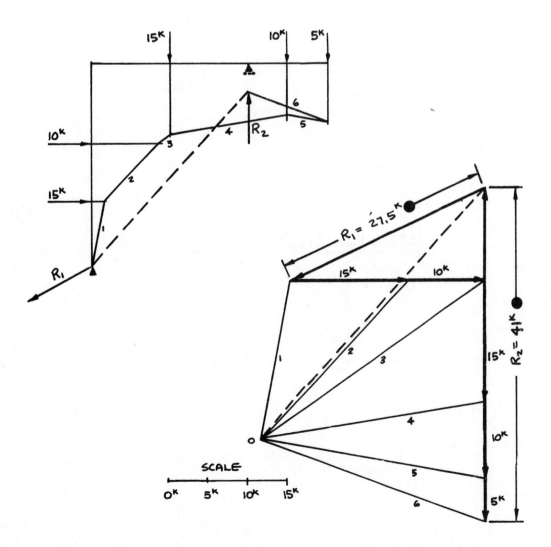

The analytical solution may be computed as a check.

$$\Sigma M_{R_1} = 0 \quad +15(8) + 10(15) + 15(10) - R_2(20) + 10(25) + 5(30) = 0$$

$$20R_2 = 120 + 150 + 150 + 250 + 150 = 820$$

$$R_2 = \frac{820}{20} = 41^k$$

$$\Sigma F_y = 0 \quad +41 - 15 - 10 - 5 - R_{1_y} = 0 \qquad R_{1_y} = -11^k$$

$$\Sigma F_x = 0 \quad +10 + 15 - R_{1_x} = 0 \qquad R_{1_x} = -25^k$$

$$R_1 = \sqrt{11^2 + 25^2} = 27.3^k$$

MECH OF MTLS 12

Given the beam shown below, with a uniform load per unit length w,
a length L, and moment of inertia I.

Solve the equation $EI \dfrac{d^2y}{dx^2} = M$ to find the equation that describes the

deflection of the beam.   M = bending moment, E = modulus of elasticity.

Due to the symmetrical loading the reactions are each equal to half the
total load.

$$R_L = R_R = \frac{Lw}{2}$$

Thus

$$M = \frac{Lw}{2}(x) - wx \frac{x}{2} = \frac{Lw}{2}x - \frac{w}{2}x^2$$

and the given equation becomes

$$EI \frac{d^2y}{dx^2} = \frac{Lw}{2}x - \frac{w}{2}x^2$$

Integrating we obtain

$$EI \frac{dy}{dx} = \frac{Lw}{4}x^2 - \frac{w}{6}x^3 + C_1 \qquad (1)$$

Because of symmetry the maximum deflection occurs at the midpoint of the span.
Thus $\frac{dy}{dx}$ (slope of the elastic curve) is equal to zero when $x = \frac{L}{2}$

Evaluating C_1

$$0 = \frac{Lw}{4}\frac{L^2}{4} - \frac{w}{6}\frac{L^3}{8} + C_1 \qquad C_1 = -\frac{wL^3}{24}$$

Equation (1) becomes

$$EI\frac{dy}{dx} = \frac{Lw}{4}x^2 - \frac{w}{6}x^3 - \frac{wL^3}{24}$$

Integrating gives us

$$EIy = \frac{wL}{12}x^3 - \frac{w}{24}x^4 - \frac{wL^3}{24}x + C_2 \qquad (2)$$

At $x = 0$ (the support), $y = 0$
Evaluating C_2 all terms are zero and hence $C_2 = 0$
Equation (2) becomes

$$EIy = \frac{wL}{12}x^3 - \frac{w}{24}x^4 - \frac{wL^3}{24}x \qquad (3)$$

Equation (3) is the desired equation. It can be rearranged as follows:

$$y = \frac{wx}{EI}\left(\frac{Lx^2}{12} - \frac{x^3}{24} - \frac{L^3}{24}\right) \; \bullet$$

MECH OF MTLS 13

A steel cylinder surrounded by a copper tube is subject to compression between parallel plates as shown:

$d = 4''$ $\qquad\qquad E_s = 30 \times 10^6$ psi

$D = 8''$ $\qquad\qquad E_c = 16 \times 10^6$ psi

$P = 100,000$ lb

REQUIRED: Determine the stresses in the copper and the steel.

Cross-Sectional Areas
Area Steel
$$A_s = \frac{\pi}{4}d^2 = \frac{\pi}{4}4^2 = 4\pi \text{ in}^2$$

Area Copper
$$A_c = \frac{\pi}{4}D^2 - A_s = \frac{\pi}{4}8^2 - 4\pi = 12\pi \text{ in}^2$$

Since the copper section and the steel section take different loads

$$P = P_c + P_s = 100,000 \text{ lbs}$$

From Hooke's law the deformation is

$$\Delta = \frac{PL}{AE} \quad \text{and from the geometry of the problem} \quad \Delta_s = \Delta_c$$

Therefore

$$\frac{P_c L}{A_c E_c} = \frac{P_s L}{A_s E_s}$$

Solving for P_c

$$P_c = P_s \left(\frac{A_c E_c}{A_s E_s}\right) = P_s \left(\frac{12\,\pi}{4\,\pi} \cdot \frac{16 \times 10^6}{30 \times 10^6}\right) = 1.6\,P_s$$

Thus

$$100,000 = P_c + P_s = 1.6\,P_s + P_s = 2.6\,P_s$$

$$P_s = \frac{100,000}{2.6} = 38,500 \text{ lbs}$$

and $\quad P_c = 100,000 - 38,500 = 61,500 \text{ lbs}$

$S = \dfrac{P}{A}$

$$S_s = \frac{P_s}{A_s} = \frac{38,500}{4\,\pi} = 3,060 \text{ psi} \; \bullet$$

$$S_c = \frac{P_c}{A_c} = \frac{61,500}{12\,\pi} = 1,630 \text{ psi} \; \bullet$$

MECH OF MTLS 14

Principal stresses occur on those planes

    (a) where the shearing stress is zero
    (b) which are 45 degrees apart
    (c) where the shearing stress is a maximum
    (d) which are subjected only to tension
    (e) which are subjected only to compression

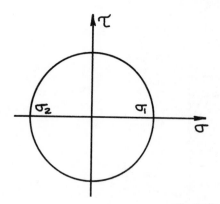

From the Mohr Circle to the left one can see that the shear at the plane of principal stresses is zero. $\bullet$

Answer is (a)

MECH OF MTLS 15

The ratio of the moment of inertia of the cross section of a beam to the section modulus is

      (a)   equal to the radius of gyration
      (b)   equal to the area of the cross section
      (c)   a measure of distance
      (d)   multiplied by the bending moment to determine
           the stress
      (e)   dependent on the modulus of elasticity of the
           beam material

I = Moment of Inertia
S = Section Modulus = I/c where c is a distance

$$\frac{I}{S} = \frac{I}{I/c} = c$$

Answer is (c) ●

MECH OF MTLS 16

When an air entrainment agent is introduced into a concrete mix

      (a)   the strength will increase
      (b)   the strength will decrease
      (c)   the strength will not be affected
      (d)   the water/cement ratio must be reduced
           from 10-15%
      (e)   calcium chloride should also be added to
           retard the time of set

Answer is (b) ●

MECH OF MTLS 17

A simple cantilever beam is 10 feet long and has a concentrated load of 100 lbs at its free end. The deflection at the free end is 1".
If the same beam is extended to 20 feet, and the load is reduced to 50 lbs, what will be the deflection at the free end? (Ignore weight of the beam.)

The Second Area-Moment Proposition tells us that the vertical displacement of the free end of a cantilever beam (B) with respect to the fixed end (A) equals the moment (with respect to B) of the area of the bending moment diagram divided by EI.

Or $= \dfrac{\text{(Area of M diagram)}}{EI} \overline{X}$

Thus

$$\Delta = \frac{1000}{EI}(10)\left(\frac{2}{3} \times 10\right) \quad \text{and} \quad \Delta' = \frac{1000}{EI}(20)\left(\frac{2}{3} \times 20\right)$$

We can see that Δ' is four times that of Δ.

Since $\Delta = 1"$, then $\Delta' = 4"$ ●

MECH OF MTLS 18

The free body of a certain beam is shown with the values given.
Draw the moment and shear diagrams for this beam, and locate the point of maximum moment.

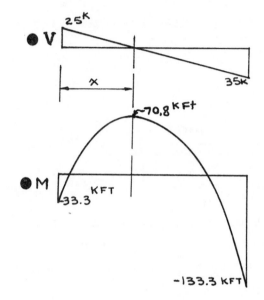

$$X = \frac{25^k}{25^k + 35^k}(20') = 8.33' \ \bullet$$

Maximum Moment
Located at point of zero shear
8.33 feet from left support

$$M_{max} = -33.3 + 25(8.33)$$
$$-3^k\left(\frac{8.33^2}{2}\right)$$

$$M_{max} = -33.3 + 208.3 - 104.2$$

$$= 70.8 \text{ k-ft}$$

MECH OF MTLS 19

Four weights of 1, 2, 3, and 4 lbs are attached to a homogeneous wheel of radius 12" as shown. Through what angle Θ should the wheel be rotated from the position shown for the weights to

    (a)   exert a maximum moment about the center of the wheel, and

    (b)   exert no moment about the center of the wheel?

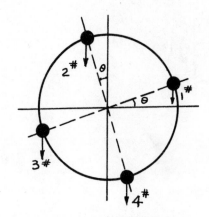

(a) $M = 2(r \sin\theta) + 3(r \cos\theta) - 4(r \sin\theta) - 1(r \cos\theta)$

   $= 2(r \cos\theta) - 2(r \sin\theta)$

$\dfrac{dM}{d\theta} = 2r(-\sin\theta - \cos\theta) = 0$

$\sin\theta = -\cos\theta$ $\tan\theta = -1$ $\theta = 135°$ or $315°$ ●

   $\theta = 135°$ gives maximum clockwise moment

   $\theta = 315°$ gives max counterclockwise moment

(b)

   $M = 2(r \cos\theta) - 2(r \sin\theta) = 0$

   $\cos\theta = \sin\theta$ $\tan\theta = 1$ $\theta = 45°$ or $225°$ ●

MECH OF MTLS 20

A contractor is to construct a water tank, a section of which is shown below, for temporary water supply. The vertical beams are in pairs, as shown, and are spaced 3 feet apart along the entire length of the tank.

Assume that the section shown is near the center of the tank, and neglect end reactions.

Draw the shear and moment diagram, and compute the reactions at the top and bottom of the tank, for one of the vertical beams shown.

Reactions

$$w = 10(62.4)(3') = 1872 \text{ lbs/ft}$$

$$F = 1872(\tfrac{10}{2}) = 9360 \text{ lbs} = 9.36^k$$

$$\Sigma M_{R_1} = 0$$

$$10R_2 - \tfrac{2}{3}(10)(9.36) = 0 \qquad R_2 = 6.24^k \; \bullet$$

$$\Sigma F_x = 0$$

$$6.24 - 9.36 + R_1 = 0 \qquad R_1 = 3.12^k \; \bullet$$

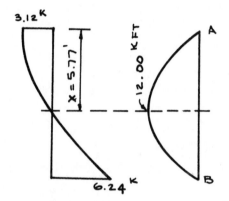

shear ●   moment ●

Compute location of zero shear and maximum moment

$$\Sigma F_x = 0 \qquad 3.12 - \tfrac{x}{10}(1.872)(\tfrac{x}{2}) = 0$$

$$x^2 = \frac{3.12(20)}{1.872} = 33.3 \qquad x = 5.77 \text{ feet}$$

Maximum Moment

$$M_{max} = 3.12(5.77) - \tfrac{5.77}{10}(1.872)(\tfrac{5.77}{2})(\tfrac{5.77}{3}) = 12.00 \text{ kip-ft}$$

MECH OF MTLS 21

A steam line has the dimensions as shown below at $70°F$. What is the total distance the end "P" will move when steam at $450°F$ is flowing in the line? (Coef. of expansion = $0.0000065 \; \frac{ft}{ft°F}$) Assume that all the distances are measured along the principal coordinate (x, y & z) axes.

x movement = $(25' + 35' + 40')(450° - 70°)(6.5 \times 10^{-6})$

$= (100)(380)(6.5 \times 10^{-6}) = 0.247$ feet

y movement = 50/100 of x movement = 0.123 feet

z movement = 10/100 of x movement = 0.025 feet

Resulting movement = $\sqrt{x^2 + y^2 + z^2} = \sqrt{0.247^2 + 0.123^2 + 0.025^2}$

$= 0.277$ feet ●

Note that this solution assumes perfect conditions; frictionless supports.

MECH OF MTLS 22

Structural steel elements subjected to torsion develop

    (a) tensile stress
    (b) compressive stress
    (c) shearing stress
    (d) moment
    (e) bending stress

    Answer is (c) ●

MECH OF MTLS 23

The deflection of a beam is

    (a) directly proportional to the modulus of elasticity and moment of inertia
    (b) inversely proportional to the modulus of elasticity and length of the beam cubed
    (c) inversely proportional to the modulus of elasticity and moment of inertia
    (d) directly proportional to the load imposed and inversely to the length squared
    (e) inversely proportional to the weight and length

The equation for the second derivative of a deflected beam is

$$\frac{d^2y}{dx^2} = \frac{M}{EI}$$

where   y = deflection at any point x from the origin
M = moment at that point
E = modulus of elasticity
I = moment of inertia

So

$$y = \iint \frac{M}{EI} \, dx \, dx$$

Thus the deflection of a beam is inversely proportional to the modulus of elasticity and moment of inertia. ●

    Answer is (c)

MECH OF MTLS 24

The differential of the shear equation is which one of the following?

    (a) load on the beam
    (b) tensile strength of beam
    (c) bending moment of the beam
    (d) slope of the elastic curve
    (e) deflection of the elastic curve

    Answer is (a) ●

MECH OF MTLS 25

A homogenous round bar of diameter D, length L, and total weight W is hung vertically from one end. If E is the modulus of elasticity, what is the total elongation of the bar?

(a) $\dfrac{WL}{\pi D^2 E}$

(b) $\dfrac{2WL}{\pi D^2 E}$

(c) $\dfrac{WL}{2\pi D^2 E}$

(d) $\dfrac{2WLE}{\pi D^2}$

(e) $\dfrac{WL}{2E}$

$$dP = dm = dy \cdot \frac{W}{L}$$

$$\Delta = \frac{PL}{AE}$$

$$= \int_0^L \frac{\frac{W}{L} dy(L - y)}{AE} = \frac{W}{ELA} \int_0^L (L - y)dy$$

$$= \frac{W}{ELA} \left[-\frac{1}{2}(L - y)^2 \right]_0^L = \frac{W}{AEL} \cdot \frac{1}{2} L^2$$

$$= \frac{WL}{2E} \cdot \frac{1}{\frac{\pi}{4} D^2} = \frac{2WL}{\pi D^2 E}$$

Answer is (b)

MECH OF MTLS 26

A vertically loaded beam, fixed at one end and simply supported at the other is indeterminate to what degree?

(a) first
(b) second
(c) third
(d) fourth
(e) statically determinate

There are three unknowns: R_1, R_2, and M

and only two equations

$$\Sigma F_v = 0 \quad \text{and} \quad \Sigma M = 0$$

Therefore the beam is indeterminate to the first degree. ●

Answer is (a)

MECH OF MTLS 27

A 60-inch diameter steel pipe, 3/8 inch thick, carries water under a pressure head of 550 feet. Determine the hoop stress in the steel.

Internal Pressure (R) = 550'(62.4 lbs/ft^3) = 34,320 lbs/ft^2 = 238 psi

Examine a section of pipe 1" long

$$2F = RA$$
$$= 238 \text{ psi } (60" \times 1")$$
$$F = 7,140 \text{ lbs}$$

$$\text{hoop stress} = \frac{F}{A} = \frac{7,140}{1" \times \frac{3"}{8}} = 19,040 \text{ psi} \ \bullet$$

MECH OF MTLS 28

Given the steel box section shown in the figure.

REQUIRED: (a) Determine the dimensions of a square, solid wood post that has the same column strength. Buckling is not a problem.

  Allowable stresses:
  wood = 400 psi
  steel = 20,000 psi

(b) If wood costs $2.75/cubic foot, installed, and steel $0.30/pound, installed, which is the more economical to use. Steel weighs 490 pounds/cubic foot.

(a) Since buckling is not a problem, the direct stress P/A governs.

$$A_s = 2(3 \times \tfrac{1}{4}) + 2(2\tfrac{1}{2} \times \tfrac{1}{4}) = 2.75 \text{ in}^2$$

Required $A_w = \dfrac{20,000}{400} \times 2.75 = 137.5 \text{ in}^2$

Width of side of square wood post = $(137.5)^{1/2}$ = 11.7 inches $\bullet$

(b) Cost/foot of steel column = $\dfrac{2.75}{144} \times 490 \times 0.30 = \$2.81$

Cost/foot of wood column = $\dfrac{137.5}{144} \times \$2.75 = \$2.63$

The installed cost of the wood column is less than that of the steel column. $\bullet$

MECH OF MTLS 29

A system consisting of two prismatical bars of equal length L and equal cross-sectional area A carries a vertical load P as shown:

Required: Determine the vertical displacement of hinge B, assuming A and C are hinged points.

Solve for F

$$F \sin 30° + F \sin 30° = P$$
$$1/2 \, F + 1/2 \, F = P$$
$$F = P$$
and $u = 1$

Solution by the Method of Virtual Work:

Bar	Length	Area	L/A	Stress due to applied load (S)	Stress due to dummy unit load (u)	$\dfrac{SLu}{A}$
AB	L	A	L/A	P	1	PL/A
BC	L	A	L/A	P	1	PL/A
						2PL/A

$$\Delta_B = \sum\left(\frac{SLu}{AE}\right) = \frac{2PL}{AE} \quad \bullet$$

MECH OF MTLS 30

Given an 18 WF 50 beam fixed at one end, pinned at the other end, and externally loaded as shown. Determine the total end moments and reactions (neglect the weight of the beam).

The restrained beam is statically indeterminate.
The solution will be obtained by the Area-Moment Method.

$$\Delta_1 = \frac{1}{EI}(21 \times 9.5)\left(\frac{9.5}{2}\right)\left(\frac{2}{3} \times 9.5\right) = \frac{7 \times 9.5^3}{EI}$$

$$\Delta_2 = 9.5\Theta = \frac{9.5}{EI}(21 \times 9.5)\left(\frac{9.5}{2}\right) = \frac{10.5 \times 9.5^3}{EI}$$

$$\Delta_3 = \frac{1}{EI}(19R_1)\left(\frac{19}{2}\right)\left(\frac{2}{3} \times 19\right) = \frac{R_1}{3EI}(2^3)(9.5^3)$$

$$\Delta_1 + \Delta_2 = \Delta_3$$

$$\frac{7 \times 9.5^3}{EI} + \frac{10.5 \times 9.5^3}{EI} = \frac{8R_1 \times 9.5^3}{3EI}$$

$$7 + 10.5 = \frac{8}{3}R_1$$

$$R_1 = \frac{3 \times 17.5}{8} = 6.56^k \;\bullet$$

$$R_2 = 21^k - R_1 = 21^k - 6.56^k = 14.44^k \;\bullet$$

$$M_A = 21^k \times 9.5' - 6.56^k \times 19'$$
$$= 199.5 - 124.6$$
$$= 74.9^{k'} \;\bullet$$

$$M_B = 0 \qquad \bullet$$

MECH OF MTLS 31

Under the loading situation and conditions shown, find the stress in each circular rod due to the 10,000 pound load. The lengths of the 3 rods were the same before the load was applied, and their deformations are equal.

$E_s = 30 \times 10^6$ psi $\qquad\qquad E_c = 18 \times 10^6$ psi

Due to the symmetry the 2 copper rods each carry the same load.

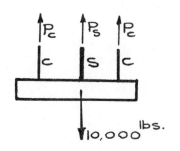

$$2P_c + P_s = 10,000 \text{ pounds} \qquad (1)$$

$$\Delta_c = \frac{P_c \, L_c}{A_c \, E_c} = \frac{P_c \times 12}{1^2 \times \pi \times 18 \times 10^6}$$

$$\Delta_s = \frac{P_s \, L_s}{A_s \, E_s} = \frac{P_s \times 12}{2^2 \times \pi \times 30 \times 10^6}$$

$$\Delta_c = \Delta_s$$

$$\frac{12P_c}{18\,\pi \times 10^6} = \frac{12P_s}{120\,\pi \times 10^6}$$

$$120P_c = 18P_s \quad \text{ or } \quad 20P_c = 3P_s \qquad (2)$$

Solving (1) and (2) gives

$$20P_c = 3(10,000 - 2P_c)$$

$$P_c = \frac{30,000}{20 + 6} = 1155 \text{ pounds}$$

$$\text{Stress}_c = \frac{1155}{\pi \times 1^2} = 367 \text{ psi} \; \bullet$$

$$P_s = 10,000 - 2(1155) = 7690 \text{ pounds}$$

$$\text{Stress}_s = \frac{7690}{\pi \times 2^2} = 611 \text{ psi} \; \bullet$$

MECH OF MTLS 32

A thin walled pressurized vessel consists of a right circular cylinder with flat ends. Midway between the ends the stress is greatest in what direction?

(a) Longitudinal
(b) Circumferential
(c) Radial
(d) At an angle of 45° to the longitudinal and circumferential direction.
(e) At an angle of 30° to the longitudinal and circumferential direction.

In thin walled cylinders the circumferential stress ("hoop tension") is twice the value of the longitudinal stress. ●

Answer is (b)

MECH OF MTLS 33

The bending moment at a section of a beam is derived from the

(a) Sum of the moments of all external forces on one side of the section.
(b) Difference between the moments on one side of the section and the opposite side
(c) Sum of the moments of all external forces on both sides of the section
(d) Sum of the moments of all external forces between supports
(e) Difference between the moments on one side and the couple on the other side

The bending moment at a section of a beam is the algebraic sum of the moments of all the external forces on one side of the section. ●

Answer is (a)

MECH OF MTLS 34

The stress concentration factor

(a) Is a ratio of the average stress on a section to the allowable stress
(b) Cannot be evaluated for brittle materials
(c) Is the ratio of areas involved in a sudden change of cross section
(d) Is the ratio of the maximum stress produced in a cross section to the average stress over the section

$$k = \frac{\text{maximum stress}}{\text{average stress}}$$ ●

Answer is (d)

MECH OF MTLS 35

Poisson's Ratio is the ratio of the

(a) Unit lateral deformation to the unit longitudinal deformation
(b) Unit stress to unit strain
(c) Elastic limit to proportional limit
(d) Shear strain to compressive strain
(e) Elastic limit to ultimate strength

$$\text{Poisson's Ratio} = \frac{\text{unit transverse deformation}}{\text{unit axial deformation}} \bullet$$

Answer is (a)

MECH OF MTLS 36

Hooke's Law for an isotropic homogeneous medium experiencing one-dimensional stress is

(a) Stress = E(strain)

(b) Strain = E(stress)

(c) (Force)(area) = $E \dfrac{\text{(change in length)}}{\text{(length)}}$

(d) $\dfrac{\text{Force}}{\text{Area}} = E \dfrac{\text{(length)}}{\text{(change in length)}}$

(e) Strain energy = E(internal energy)

Hooke's Law tells us that stress is proportional to strain.

Stress = Constant x Strain

The constant is called the modulus of elasticity (E).
Thus

Stress = E(Strain) ●

Answer is (a)

MECH OF MTLS 37

The modulus of rigidity of a steel member is:

(a) a function of the length and depth.
(b) defined as the unit shear stress divided by the unit shear deformation.
(c) equal to the modulus of elasticity divided by one plus Poisson's ratio.
(d) defined as the length divided by the moment of inertia.
(e) equal to approximately seven-tenths of the modulus of elasticity.

modulus of rigidity = shear modulus, G, psi = $\dfrac{S_s, \text{ psi}}{\phi, \text{ radians}}$

$G = \dfrac{\text{shear stress}}{\text{shear deformation (strain)}}$

Answer is (b) ●

MECH OF MTLS 38

A thin homogeneous metallic plate containing a hole is heated sufficiently to cause expansion. If the coefficient of surface expansion is linear, the area of the hole will:

(a) Increase at twice the rate the area of the metal increases
(b) Increase at the same rate as the area of the metal increases
(c) Stay the same
(d) Decrease at the same rate as the area of the metal increases
(e) Decrease at twice the rate the area of the metal increases

Answer is (b) ●

MECH OF MTLS 39

What is the maximum moment of a loaded beam given only the following?

SHEAR DIAGRAM

MOMENT DIAGRAM

The solution is based on two facts:
1. The maximum value of bending moment occurs where the shear passes through zero.
2. The change in bending moment between two sections of a beam equals the area of the shear diagram between those sections, taking the sign of the shear into account.

The location of zero shear (on the shear diagram) must first be determined.

Its distance from the right end of the beam $= \frac{12}{20}(25'-5') = 12$ feet

The bending moment at this point:

$$M_{max} = 12^k \times \frac{12'}{2} = +72 \text{ kip-feet} \bullet$$

MECH OF MTLS 40

A steel bar is mounted as shown, between unyielding supports. If the
temperature of the bar increases by ΔT, find the unit stress in the bar in
terms of the coefficient of thermal expansion α and the modulus of
elasticity E.

α = coefficient of thermal expansion

E = modulus of elasticity

ΔT = increase in termperature

Strain $S = \alpha(\Delta T)$

Since Stress $S = ES$ Then Stress $S = E\alpha(\Delta T)$ ●

Note that in a body in which <u>all</u> temperature deformation is prevented, the
unit stress caused by   thermal expansion is independent of the length L
of the body.

MECH OF MTLS 41

A beam of uniform cross-section, symmetrical about its center, is
subjected to pure bending. The fiber stresses in the beam vary from
zero in the center to a maximum in the extreme fibers, and the resisting
moment varies in a similar manner. I-beams and wide flange beams are
two types of beams designed to take advantage of this fact in order to
save weight.

If the rectangular beam shown is used, how many times the resisting
moment of area A (the middle two quarters) would areas B (outer two
quarters) have?

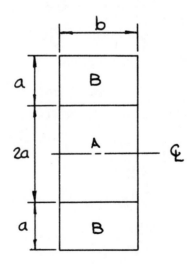

The flecture formula
$$f_s = \frac{Mc}{I}$$

$$M_{resisting} = \frac{f_s I}{c} \qquad \text{For a given section} \quad M_{resisting} \propto I$$

$$I_A = \frac{bd^3}{12} = \frac{b(2a)^3}{12} = \frac{8ba^3}{12} \qquad I_{A+B} = \frac{bd^3}{12} = \frac{b(4a)^3}{12} = \frac{64ba^3}{12}$$

$$I_B = I_{A+B} - I_A = \frac{64ba^3}{12} - \frac{8ba^3}{12} = \frac{56ba^3}{12}$$

$$\frac{M_{resisting(B)}}{M_{resisting(A)}} \propto \frac{I_B}{I_A} = \frac{\frac{56ba^3}{12}}{\frac{8ba^3}{12}} = 7 \quad \bullet$$

MECH OF MTLS 42

A steel structural ∠6"x6"x1/2" must be welded to the top of a steel **WF** beam. The design weld value for the weld shown is 3,600 lb/lin. inch.

If the tension load on the angle is 20,000 pounds, show the length and placing of welds so that no eccentricity will exist in the connection.

PLAN VIEW

Required length of weld $= \dfrac{20,000}{3,600} = 5.57$ inches

Let   X = Length of weld on the side with the outstanding leg
    5.57 - X = Length of weld on the right side of the angle

For no eccentricity the centroid of the welds must be at the center of gravity of the angle, or

$$1.68X = (6 - 1.68)(5.57 - X)$$
$$= 24 - 4.32X$$
$$6X = 24 \qquad\qquad\qquad X = 4.00 \text{ inches} \;\bullet$$
$$5.57 - X = 1.57 \text{ inches} \;\bullet$$

MECH OF MTLS 43

Determine the deflection of the uniformly loaded beam with simply supported ends, as a function of X. Let W be the load per unit length, L the length, E the modulus of elasticity, and I the moment of inertia.

Given: $M = EI \dfrac{d^2y}{dx^2}$

$$M = EI \dfrac{d^2y}{dx^2} \qquad\qquad \Sigma M_o = 0 \qquad \left(+\downarrow\right)$$

$$\dfrac{WL}{2}(X) - WX\left(\dfrac{X}{2}\right) - M = 0$$

$$M = \dfrac{WLX}{2} - \dfrac{WX^2}{2}$$

Successive integrations give:

$$EI \dfrac{dy}{dx} = \dfrac{WLX^2}{4} - \dfrac{WX^3}{6} + C_1 \qquad (1)$$

$$EIy = \dfrac{WLX^3}{12} - \dfrac{WX^4}{24} + C_1 X + C_2 \qquad (2)$$

Boundary conditions:

$$\frac{dy}{dx} = 0 \qquad \text{when} \quad X = \frac{L}{2}$$

$$y = 0 \qquad \text{when} \quad X = 0$$

Substituting the boundary conditions into (1) and (2)

$$EI(0) = \frac{WL(L/2)^2}{4} - \frac{W(L/2)^3}{6} + C_1 \qquad C_1 = -\frac{WL^3}{24}$$

$$EI(0) = \frac{WL(0)^3}{12} - \frac{W(0)^4}{24} - \frac{WL^3}{24}(0) + C_2 \qquad C_2 = 0$$

Thus equation (2) becomes

$$EIy = \frac{WLX^3}{12} - \frac{WX^4}{24} - \frac{WL^3 X}{24}$$

$$\text{Deflection } y = \frac{2WLX^3 - WX^4 - WL^3 X}{24EI} \quad \bullet$$

MECH OF MTLS 44

The maximum bending moment of a beam simply supported at both ends and subject to a total load W uniformly distributed over its length L is expressed by the formula

(a) WL/8

(b) $WL^2/8$

(c) WL/2

(d) $WL^2/2$

(e) WL/4

$$W = \omega L$$

$$\sum M_o = 0 \quad \left(\overset{\curvearrowleft}{+}\right) \qquad \left(\frac{W}{2} \times \frac{L}{2}\right) - \left(\frac{W}{2} \times \frac{L}{4}\right) - M = 0$$

$$M = \frac{WL}{4} - \frac{WL}{8} = \frac{WL}{8} \quad \bullet$$

Answer is (a)

MECH OF MTLS 45

The maximum moment induced in a simply supported beam of 20 foot span, by a 2,000 pound load at midspan is

    (a)   125,000 ft-lbs
    (b)   15,000 ft-lbs
    (c)   30,000 ft-lbs
    (d)   10,000 ft-lbs
    (e)   7,500 ft-lbs

$$M_A = 10(1,000) = 10,000 \text{ lb-ft} \; \bullet$$

Answer is (d)

MECH OF MTLS 46

The three moment equation may be used to analyze

    (a)   a continuous beam.
    (b)   a beam loaded at the third points.
    (c)   a step tapered column.
    (d)   a three element composite beam.
    (e)   an axially end-loaded beam.

Answer is (a) ●

MECH OF MTLS 47

A thin hollow sphere of radius 10" and thickness 0.1" is subjected to an internal pressure of 100 psig. The maximum normal stress on an element of the sphere is

    (a)   5,000 psi
    (b)   7,070 psi
    (c)   10,000 psi
    (d)   14,140 psi
    (e)   20,000 psi

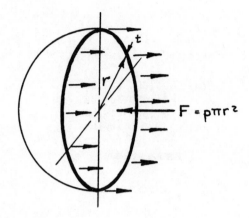

Unit pressure = p

Cross-sectional area of wall of sphere = $2\pi rt$

Area of cross-section through the sphere = πr^2

Force F = unit pressure x cross-sectional area over which it acts = $p\pi r^2$

$$\text{Stress} = \frac{\text{Force F}}{\text{Area of wall of sphere}} = \frac{p\pi r^2}{2\pi rt} = \frac{pr}{2t} = \frac{100(10)}{2(0.1)} = 5,000 \text{ psi} \bullet$$

Answer is (a)

MECH OF MTLS 48

A tension member   made up of two angles carries a load of 160 kips as shown. The angles must be welded to a 7/8" gusset plate with the 4" leg of the angles outstanding. If the value of the weld is 3000 lbs/inch, design the lengths and placement of the welds to suit the given condition.

Load per angle = 80^k

Inches of weld per angle = $\dfrac{80^k}{3^k/\text{inch}}$ = 26.7 inches

We want the resultant force exerted by the welds to be collinear with the 80 kip force at the centroid of the angle.

$$d_1 = 2.47 \text{ inches}$$
$$d_2 = 8 - 2.47 = 5.53 \text{ inches}$$

$$2.47 L_1 - 5.53 L_2 = 0$$
$$L_1 + L_2 = 26.7$$

Solving the two equations:

$$L_1 = 26.7 - L_2$$
$$2.47(26.7 - L_2) - 5.53 L_2 = 0$$
$$66 - 2.47 L_2 - 5.53 L_2 = 0$$

$$-8 L_2 = -66 \qquad\qquad L_2 = 8.25 \text{ inches} \ \bullet$$
$$L_1 = 26.7 - 8.25 \qquad\qquad L_1 = 18.45 \text{ inches} \ \bullet$$

MECH OF MTLS 49

Calculate the reactions at A and B and draw the shear and moment diagrams. Neglect the weight of the beam.

$$\Sigma M_B = 0 \qquad -8(500) + 20 R_A - 23(200 \times 6) + 1{,}000 = 0$$
$$20 R_A = 4000 + 23(1{,}200) - 1{,}000$$
$$R_A = \frac{3{,}000 + 27{,}600}{20} = 1{,}530 \text{ lbs} \ \bullet$$

$$\Sigma M_A = 0 \qquad 1{,}000 - 3(200 \times 6) + 12(500) + 20(1{,}000) - 20 R_B = 0$$
$$20 R_B = 1{,}000 - 3{,}600 + 6{,}000 + 20{,}000$$
$$R_B = \frac{23{,}400}{20} = 1{,}170 \text{ lbs} \ \bullet$$

SHEAR DIAGRAM ●

MOMENT DIAGRAM ●

MECH OF MTLS 50

Force F, and moment M, act at point P. Both bars have length L, Area A, modulus of elasticity E, and moment of inertia I.

(a) What is the vertical deflection of Q in terms of the parameters given above?
(b) What is the horizontal deflection of Q in terms of the parameters given above?

Solution by the conjugate beam method:

MOMENT DIAGRAM

$$Q_H = \left[\left(\frac{FL^2}{2}\right)\left(\frac{2L}{3}\right) + (ML)\left(\frac{L}{2}\right) \right] \frac{1}{EI} = \frac{FL^3}{3EI} + \frac{ML^2}{2EI} \longrightarrow \bullet$$

$$Q_V = \left[\left(\frac{FL^2}{2}\right)(L) + (ML)(L) \right] \frac{1}{EI} = \frac{FL^3}{2EI} + \frac{ML^2}{EI} \downarrow \bullet$$

MECH OF MTLS 51

Determine the maximum tensile bending stresses and compressive bending stresses in the cantilever T-beam shown below.

$\text{Stress} = \frac{Mc}{I}$ The maximum stresses will occur at the cross-section where the moment M is a maximum. For this cantilever the maximum moment is at the face of the wall

$M_{max} = (500 \text{ lb/ft})(8 \text{ ft})(4 \text{ ft}) = 16,000 \text{ lb-ft}$

$$\bar{X} = \frac{\Sigma A \bar{X}}{\Sigma A} = \frac{(8)(1)(7.5) + (7)(1)(3.5)}{15} = \frac{84.5}{15} = 5.63 \text{ inches}$$

Moment of inertia for composite area:

For rectangular section: $I_o = \frac{bd^3}{12}$

$I_x = I_o + Ad^2$

$I_x = \frac{8 \times 1^3}{12} + 8(7.5 - 5.63)^2 + \frac{1 \times 7^3}{12} + 7(5.63 - 3.5)^2$

$\quad = 0.67 + 28.0 + 28.58 + 31.75 = 89.0 \text{ in}^4$

BENDING STRESS DIAGRAM

Maximum compressive stress $= \frac{Mc}{I} = \frac{(16,000 \text{ ft-lb})(12 \text{ in/ft})(5.63 \text{ in})}{89.0}$

$\qquad = 12,140 \text{ psi} \; \bullet$

Maximum tensile stress $= \frac{Mc}{I} = \frac{(16,000 \text{ ft-lb})(12 \text{ in/ft})(2.37 \text{ in})}{89.0}$

$\qquad = 5,120 \text{ psi} \; \bullet$

MECH OF MTLS 52

A 20" long bimetallic strip is made of metal A and metal B, each 0.020 inches thick. Assuming that the strip is straight at a temperature of 68°F, and is fixed at one end and free to move at the other, compute the deflection of the free end at a temperature of 86°F, if the curvature of the strip is in the form of a circle.

The linear coefficient of expansion for metal A is 24×10^{-6} inches/inch/°C

The linear coefficient of expansion for metal B is 14×10^{-6} inches/inch/°C

Assume that both metals have the same modulus of elasticity.

Determine the temperature change in degrees Centigrade:

$$°C = \frac{5}{9} (F - 32) \qquad °C = \frac{5}{9} (68 - 32) = 20°C$$

$$°C = \frac{5}{9} (86 - 32) = 30°C$$

Difference in expansion between A and B

$$= (24 \times 10^{-6} - 14 \times 10^{-6})(30°C - 20°C)(20 \text{ inches}) = 0.002$$

The shape assumed by the bimetallic strip will be a circular segment.

The circumference of a circle is $2\pi R$

The circumferential length of the circular segment (S) = $R\Theta$ for any angle Θ (in radians).

Let L = heated length of metal B
   R = radius to centerline of the bimetallic strip

Substituting into $S = R\Theta$

Metal B	$L = (R - 0.01)\Theta$	(1)
Metal A	$L + 0.002 = (R + 0.01)\Theta$	(2)

Substituting the value of L from (1) into (2)

$$(R - 0.01)\Theta + 0.002 = (R + 0.01)\Theta$$

$$R\Theta - 0.01\Theta + 0.002 = R\Theta + 0.01\Theta$$

$$\Theta = \frac{0.002}{0.02} = 0.1 \text{ radian} = 5.73 \text{ degrees}$$

$$R = \frac{S}{\Theta} = \frac{20}{0.1} = 200 \text{ inches}$$

Deflection = $R - R \cos \Theta$

= 200 - 200(0.995)

= 200 - 199 = 1 inch ●

MECH OF MTLS 53

Fixing both ends of a simply supported beam that has only a concentrated load at midspan will increase the allowable load by:

        (a)   25%
        (b)   50%
        (c)   100%
        (d)   200%
        (e)   no change

SHEAR DIAGRAM   SHEAR DIAGRAM

MOMENT DIAGRAM   MOMENT DIAGRAM

$$R_A = R_B = \frac{P}{2}$$

$$M_{max} = \frac{P}{2} \cdot \frac{L}{2} = \frac{PL}{4}$$

$$R_A = R_B = \frac{P}{2}$$

$$M_{max} = \frac{P}{2} \cdot \frac{L}{4} = \frac{PL}{8}$$

Based on maximum moment, the beam with fixed ends will support twice the applied load of the simple beam. ⬤

Answer is (c)

MECH OF MTLS 54

In a long column (slenderness ratio > 160) which of the following has the greatest influence on its tendency to buckle under a compressive load.

        (a)   The modulus of elasticity of the material
        (b)   The compressive strength of the material
        (c)   The radius of gyration of the column
        (d)   The length of the column
        (e)   The moment of inertia of the column

Euler's formula for slender columns:

$$P_{crit} = \frac{\pi^2 EI}{L^2}$$

Answer is (d) ⬤

MECH OF MTLS 55

The area of the shear diagram of a beam between any two points on the beam is equal to the

(a) Change in shear between the two points
(b) Total shear beyond the two points
(c) Average moment between the two points
(d) Change in moment between the two points
(e) Change in deflection between the two points.

Answer is (d) ●

MECH OF MTLS 56

Poisson's ratio is principally used in

(a) The determination of capability of a material for being shaped
(b) The determination of capacity of a material for plastic deformation without fracture
(c) Stress-strain relationships where stresses are applied in more than one direction
(d) The determination of the modulus of toughness
(e) The determination of the endurance limit

$$\text{Poisson's ratio } \mu = \frac{\text{unit transverse deformation}}{\text{unit axial deformation}}$$

Answer is (c) ●

MECH OF MTLS 57

A railroad track is laid at a temperature of $15°F$ with gaps of 0.01 feet between the ends of the rails. The rails are 33 feet long. If they are prevented from buckling, what stress will result from a temperature of $110°F$?

$$\Delta L = \alpha L \Delta T \quad \text{where} \quad \alpha = 6.5 \times 10^{-6} \text{ per } °F$$
$$L = 33 \text{ feet}$$
$$\Delta T = 110 - 15 = 95°$$

$\Delta L = 6.5 \times 10^{-6} \times 33 \times 95 = 0.0204$ feet. This elongation cannot occur and hence there is a stress induced in the rail.

The strain in the rail = 0.0204 - 0.01 = 0.0104 ft/33 ft rail

$$= 0.00031 \text{ ft/ft (or in/in)}$$

The thermally induced compressive stress:

$$S = \epsilon_t E = 0.00031 \times 30 \times 10^6 = 9300 \text{ psi} \ ●$$

MECH OF MTLS 58

The structure shown on the next page is pinned at A and rests on friction-less rollers at B. The structure has equal legs of length L. If both legs are loaded by a triangular load distribution of maximum value w lb/ft, determine:

(a) The reactions at points A and B.
(b) At what point in the structure is the bending moment a maximum?
(c) What is the magnitude of this maximum bending moment?

Due to the frictionless rollers at B, $V_B = 0$ ●

$\sum M_A = 0$ ↻ (+) $\quad \frac{wL}{2}(\frac{2}{3}L) + \frac{wL}{2}(\frac{2}{3}L) - H_B(L) = 0$

$$\frac{wL^2}{3} + \frac{wL^2}{3} - LH_B = 0 \qquad H_B = \frac{2wL}{3} \longleftarrow \;\bullet$$

$\sum F_H = 0$

$$\frac{wL}{2} - \frac{2wL}{3} + H_A = 0 \qquad H_A = \frac{wL}{6} \longrightarrow \;\bullet$$

$\sum F_V = 0$

$$V_A - \frac{wL}{2} = 0 \qquad\qquad V_A = \frac{wL}{2} \uparrow \;\bullet$$

By inspection we see that the maximum bending moment occurs at C ●

$$M_{max} = M_C = \frac{wL}{2} \cdot \frac{2}{3}L = \frac{wL^2}{3} \;\bullet$$

5
MATERIALS SCIENCE

MAT SCI 1

Young's modulus of elasticity for a material can be calculated indirectly from which of the following properties of the material?

    (a) Temperature coefficient of expansion and dielectric constant.

    (b) Temperature coefficient of expansion and specific heat.

    (c) Density and velocity of sound in the material.

    (d) Density and interatomic spacing in the material.

    (e) Coefficient of conductivity for heat and for electricity.

Speed of longitudinal compressional sound waves in ft/sec is dependent on modulus of elasticity E, lb/in^2. Dimensional comparisons among variables assist in making the proper selection:

    (a) α, $({}^\circ F^{-1})$ and k, (dimensionless)

    (b) α, $({}^\circ F^{-1})$ and C_p, $(BTU/lb_m\,{}^\circ F)$

    (c) ρ, $(slug/ft^3)$ and V, (ft/sec)

    (d) ρ, $(slug/ft^3)$ and ℓ, (ft)

    (e) k, $(BTU/(hr\ ft^2)({}^\circ F/ft)$ and σ, $(ohm^{-1}\ meter^{-1})$

Since slug = $lb_f sec^2/ft$, manipulation of (c) yields the relationship:

$$E,\ \frac{lb}{in^2} = \rho\left(\frac{lb_f sec^2/ft}{ft^3}\right)\ V^2\ (ft/sec)^2 \times (1\ ft^2/144\ in^2) = \rho\frac{V^2}{144}$$

If ρ = lb_m/ft^3, divide by g_c = 32.2 to convert lb_m to slugs:

$$E = \frac{\rho V^2}{(32.2)(144)}$$

Answer is (c) ●

MAT SCI 2

The linear portion of the stress-strain diagram of steel is known as the

    (a) modulus of elasticity
    (b) plastic range
    (c) irreversible range
    (d) elastic range
    (e) secant modulus

The portion of the diagram where stress is linearly proportional to strain is called the elastic range. Yield strength is usually determined at 0.2% offset, while elastic modulus is calculated from slope of the linear portion of the stress-strain diagram. Non-linear materials such as rubber, plastics, or concrete use a secant modulus of elasticity based on a line from the origin to a predetermined fraction of ultimate tensile or compressive strength.

<u>Tensile Stress - Strain Diagram for ductile steel</u>

Answer is (d) ●

MAT SCI 3

"Modulus of Resilience" is

    (a) the same as the modulus of elasticity.
    (b) a measure of a materials ability to store strain energy.
    (c) the reciprocal of the modulus of elasticity.
    (d) a measure of the deflection of a member.
    (e) a measure of the strain induced in a member due to a change in temperature.

The modulus of resilience is the amount of energy per unit of volume that a given material stores when stressed to the elastic limit. ●

Answer is (b)

MAT SCI 4

Imperfections within metallic crystal structures may be all of the following, EXCEPT:

    (a) lattice vacancies, or extra interstitial atoms.

    (b) ion pairs missing in ionic crystals (Shotky imperfections).

    (c) displacement of atoms to interstitial sites (Frankel defects).

    (d) twinning to form mirror images along a low energy boundary or crystal plane.

    (e) linear defects, or slippage dislocations caused by shear.

All are correct except (b). Metals contain atoms at the lattice sites. Salts and other ionic crystals contain ions in a charge-balanced structure at adjacent lattice sites.

The false statement is (b) ●

MAT SCI 5

All of the following statements about strain energy are correct, EXCEPT:

    (a) It is caused by generation and movement of dislocations through shear or plastic deformation.

    (b) It results from trapped vacancies in the crystal lattice.

    (c) It is proportional to length of dislocation, shear modulus, and shortest distance between equivalent lattice sites (points).

    (d) It is dependent on energy differences existent between stable regularly spaced sites and the metastable sites around point defects.

    (e) It is less for sites at grain boundaries than for internal sites within the crystal structure.

All except (e) are correct. Grain boundaries are high energy areas of mismatch where adjacent crystals come in contact.

The false statement is (e) ●

MAT SCI 6

All of the following statements about diffusion and grain growth are correct, EXCEPT:

    (a) Vacancies and interstitial atoms affect diffusion, whose net result is movement of atoms to produce a structure of less strain and of uniform composition.

    (b) Diffusion is irreversible and requires an activation energy; its rate increases exponentially with temperature. It follows the diffusion equation where flux equals diffusivity times concentration gradient.

    (c) Activation energy for diffusion through structures is inversely proportional to atomic packing factor in the lattice.

    (d) Atoms can diffuse both within crystals and across grain (crystal) boundaries.

    (e) Grain growth results from diffusion and minimizes total grain boundary area. Large grains grow at the expense of small ones, and grain boundaries move towards their centers of curvature.

All except (c) are correct. When atoms move from one site to another, bonds are broken and reconstituted. During transition an activation energy is required for distortion of the lattice. Small solute atoms, lower melting point solvent, and lower atomic packing factor in the lattice all reduce activation energy. Hence, activation energy for diffusion is <u>directly</u> proportional to packing factor.

The false statement is (c) ●

MAT SCI 7

All of the following statements about slip are correct, EXCEPT:

  (a) Slip, or shear along crystal planes, results in an irreversible plastic deformation or permanent set.

  (b) It involves only a few atoms at a time in a series of small dislocation movements.

  (c) Slip planes lie in the direction of the longest distance between neighboring sites in the crystal lattice.

  (d) Ease of slippage is directly related to number of low energy slip planes existent in the lattice structure.

  (e) Slip is impeded by solution hardening, wherein large solute atoms serve as anchor points around which slippage does not occur.

Plastic deformation usually occurs by slip. Slip planes lie in the direction of the <u>shortest</u> distance (Burgess vector) between neighboring sites in the crystal lattice.

The false statement is (c) ●

MAT SCI 8

When a metal is cold worked all of the following generally occur EXCEPT:

  (a) Recrystallization temperature decreases
  (b) Ductility decreases
  (c) Grains become equi-axed
  (d) Tensile strength increases
  (e) Slip or twinning occurs.

Cold working, such as rolling, forging, drawing, or extrusion, reduces cross sectional area at temperatures below the recrystallization temperature. Strain hardening occurs, increasing both yield and ultimate strength. Internal strains and minute cracks are introduced as slip or twinning occur. Ductility, elongation and recrystallization temperature are decreased. A preferred grain orientation is introduced in the direction of elongation, but the grains are <u>not</u> <u>equi-axed</u>.

The false statement is (c) ●

MAT SCI 9

All of the following statements about strain hardening are correct, EXCEPT:

  (a) Strain hardening strengthens metals. Resistance to deformation increases with the amount of strain present.

  (b) Strain hardening is relieved during softening, annealing above the recrystallization temperature.

  (c) Strain hardening is produced by cold working (deformation below the recrystallization temperature).

  (d) More strain hardening requires more time-temperature exposure for relief.

  (e) Strain hardening is relieved during recrystallization. Recrystallization produces less strained and more ordered structures.

All are correct except (d). Time at or above the recrystallization temperature allows relief of strain hardening. More cold working produces more strain hardening, but this also lowers recrystallization temperature and requires a <u>lesser</u> time-temperature exposure for relief.

The false statement is (d) ●

MAT SCI 10

All of the following processes strengthen metals, EXCEPT:

- (a) Annealing above the recrystallization temperature.
- (b) Work hardening by mechanical deformation below the recrystallization temperature (cold working).
- (c) Precipitation processes, such as age hardening, which produce high strength by formation of submicroscopic phases during low temperature heat treatment.
- (d) Heat treatments such as quenching and tempering, for production of a finer microstructure.
- (e) Increasing the carbon content of low carbon steels.

All except (a) are correct. Strength and hardness are increased at the expense of ductility and toughness (opposite of brittleness) by developing structures which retard dislocation movement. Annealing increases ductility, but reduces both strength and hardness by relieving strain or work hardening.

The false statement is (a) ●

MAT SCI 11

The valence band model used to explain metallic conduction is based on all of the following statements, EXCEPT:

- (a) Each valence band may contain up to 2n electrons/n atoms; each electron lies at a discretely different energy level.
- (b) Fermi energy level, E_f, is essentially temperature independent, and is the energy at which 50% of available energy states are occupied.
- (c) A conduction band lies at the next higher set of electronic energy levels above those occupied at the ground state.
- (d) An energy gap exists between the valence band and the conduction band in semiconductors and insulators.
- (e) Conduction occurs when an electron remains in its existent valence band.

All except (e) are correct. Conduction occurs when an electron-hole pair are produced by thermal energy and the electron receives sufficient energy to occupy a vacant and accessible energy state in the higher energy conduction band. In insulators the large energy gap between valence and conduction bands severely reduces the probability of an electron obtaining sufficient energy to jump the gap and to exist in the higher energy conduction band.

The false statement is (e) ●

MAT SCI 12

Intrinsic silicon becomes extrinsically conductive, with electrons as majority carriers, when doped with which of the following?

    (a) nothing
    (b) antimony
    (c) boron
    (d) germanium
    (e) aluminum

Extrinsic conduction in group IV elements (Silicon or Germanium) is produced by doping with a very small fraction of a ppm of a group III element (Boron or Aluminum) or a group V element (Arsenic or Antimony).

Group III elements have a deficiency of electrons relative to the group IV structure; they produce p-type semiconduction where holes are the majority charge carriers, and any electrons in the valence conduction band are minority carriers.

Group V elements have an excess of electrons, relative to the group IV structure; they produce n-type semiconduction where electrons in the conduction band are majority charge carriers, and holes are minority carriers.

Answer is (b) ●

MAT SCI 13

When the emitter to base of an npn transistor is forward biased, and base to collector is reverse biased, all of the following are correct EXCEPT:

    (a) Electrons are majority carriers in the n-emitter and n-collector regions.

    (b) Electrons are minority carriers in the p-base region.

    (c) Holes are majority carriers in the p-base region.

    (d) The emitter is positive with respect to the collector.

    (e) Holes are minority carriers in the n-emitter and n-collector regions.

Forward bias increases and reverse bias decreases negative charge on an n-region. Electrons are majority carriers in n-regions and are the minority carriers in p-regions. A forward biased n-type emitter acts as an electron source.

Symbolically, the emitter is shown with an arrowhead in the direction from p- to n-regions. Base is very thin to minimize the recombination of opposite moving electrons and holes, and to reduce their transit times.

In basic circuits the emitter to base is forward biased and base to collector is reverse biased. For an npn transistor (n-emitter, p-base, n-collector), the base must be positive with respect to the emitter, and the emitter must be <u>negative</u> with respect to the collector. For a pnp transistor these bias voltage polarities are reversed.

Based on the above transistor principles, all <u>except</u> (d) are correct.

The false statement is (d) ●

MAT SCI 14

All of the following statements about solid solutions are correct, EXCEPT:

(a) Solid solutions can result when basic structure of the solvent can accomodate solute additions.

(b) In solid solutions larger solute atoms occupy the interstitial space between solvent atoms that are located at the lattice sites.

(c) Solid solutions may result by substitution of one atomic species for another, provided radii and electronic structures are compatible.

(d) Order - disorder transitions that occur at elevated temperature in solid solutions involve changes due to thermal agitation from preferred orientation to random occupancy of lattice sites.

(e) Defect structures exist in solid solutions of ionic compounds because vacancies are required to maintain an overall charge balance when there are differences in oxidation state of solute and solvent ions.

All are correct except (b). Interstitial solid solutions result when very small solute atoms, such as H, C or N, occupy interstitial space between larger solvent atoms located at the lattice sites.

Example: carbon interstitially dissolved in the face centered cubic austenite (γ- iron) structure of steels.

The false statement is (b) ●

MAT SCI 15

All of the following statements about ferromagnetism are correct, EXCEPT:

(a) Magnetic domains are small volumes existent within a single crystal where atomic magnetic moments are unidirectionally aligned.

(b) Domains are randomly oriented when unmagnetized. On magnetization, domains oriented with the external field grow at the expense of unaligned domains.

(c) Impurities, inclusions and strain hardening interfere with change of domain boundaries, and add to the permanency of a magnet.

(d) High magnetic susceptibility of ferromagnetic materials disappears below the Curie temperature.

(e) Above the Curie temperature ($770°C$, $1418°F$ for iron) ferromagnetic domains disappear and the metal becomes simply paramagnetic, a consequence of unpaired electrons.

All are correct except (d). Spontaneous magnetism diminishes at elevated temperatures and vanishes above the Curie temperature.

Atomic magnetic moments (Bohr magnetons) arise mainly from unbalanced electron spins in the 3d orbitals.

The false statement is (d) ●

MAT SCI 16

magnesium-zinc phase diagram

The above phase diagram indicates which of the following intermetallic compounds?

(a) $MgZn_2$

(b) Mg_2Zn

(c) $MgZn$

(d) Mg_2Zn_3

(e) Mg_3Zn_2

Compound formation splits the phase diagram into separate diagrams to the left and right of the compound composition at 67 mole % Zn and 33 mole % Mg, or $MgZn_2$. The region ABJ represents solubility of the compound in solid magnesium, while C and F are eutectic compositions.

Intermetallic compounds are characterized by nearly perfect structural ordering among unlike neighboring atoms. They have the typical microstructure of a pure metal, an aggregate of crystals of one kind. They are usually brittle and mechanically weak.

Answer is (a) ●

MAT SCI 17

Refer to the magnesium - zinc phase diagram of the previous question. At 575°C a liquid of eutectic composition C (29 mole % Zinc) has sufficient zinc added to raise zinc composition to 40 mole %. When this new liquid is cooled, the first solid phase to separate is:

(a) solid solution containing less than 1% intermetallic compound dissolved in Mg.

(b) solid intermetallic compound

(c) a mixture of solid intermetallic compound and solid eutectic C (29 mole % Zn)

(d) solid eutectic C (29 mole % Zn)

(e) a mixture of solid intermetallic compound and solid eutectic F (92 mole % Zn)

〰〰〰〰〰〰〰〰

On cooling, the first phase transition occurs on line C-I of the diagram at about 419°C, here much liquid (40 mole % Zn along CI) and a slight amount of solid intermetallic compound (67 mole % Zn along IE) exist. On further cooling to just above 345°C the system contains $\frac{67 - 40}{67 - 29}$ mole fraction of liquid eutectic C (29 mole % Zn) and $\frac{40 - 29}{67 - 29}$ mole fraction of solid intermetallic compound D (67 mole % Zn), computed by tie line lever rule. Further cooling yields solid eutectic C and solid intermetallic compound D at the calculated mole fractions.

Answer is (b) ●

MAT SCI 18

All of the following statements about steels are correct, EXCEPT:

(a) Yield strength of commercially available heat treated alloy steels does not exceed 200,000 psi.

(b) High temperature alloys used in jet engine turbine blades can withstand 2000°F continuously over extended periods.

(c) Abrasion resistance of extra strength steels may be obtained by increasing hardness to 225 - 400 Brinell at the expense of some ductility and toughness.

(d) Intergranular corrosion of chromium - nickel stainless steels is reduced when stabilized by addition of columbium (niobium), titanium or tantalum to preferentially form carbides and prevent chromium depletion and chromium carbide precipitation in grain boundary areas.

(e) Steels used in reinforced concrete construction include ASTM A-15 billet and A-432 high strength billet with tensile yield strengths from 33,000 to 75,000 psi.

〰〰〰〰〰〰〰〰

Over 100 steels are available with yield strengths above 225,000 psi. Any steel with yield strength above 160,000 psi is considered an ultra-high strength steel.

Examples are: medium carbon quenched and tempered alloy steels
medium alloy air hardening steels
martensitic stainless steels
cold rolled austenitic stainless steels
semi-austenitic stainless steels
high alloy quenched and tempered steels
maraging steels
cold worked plain high carbon steels
The last two types can have ultimate strengths in excess of 300,000 psi.

The false statement is (a) ●

MAT SCI 19

IRON-IRON CARBIDE PHASE DIAGRAM

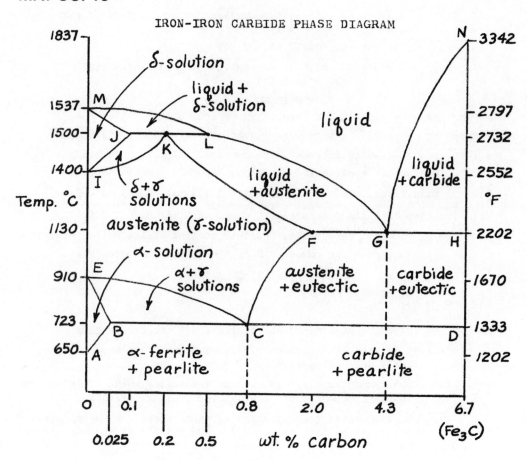

On the usual iron - iron carbide phase diagram above, all of the following are true, EXCEPT:

(a) a eutectic reaction occurs at G, 1130°C.

(b) a peritectic reaction occurs at K, 1500°C.

(c) a eutectoid solid reaction occurs at C, 723°C.

(d) a peritectoid solid reaction occurs at F, 1130°C.

(e) the eutectoid composition is at 0.8 wt.% C.

〰〰〰〰

All are correct except (d). Typical phase reactions involve transition between one phase and two phases. As outlined below, 1, 2 and 3 denote increasing wt.% composition:

(1) eutectic reaction:

$$Liq_2 \xrightleftharpoons[\text{heat}]{\text{cool}} Solid_1 + Solid_3$$

(2) peritectic reaction:

$$Solid_1 + Liq_3 \xrightleftharpoons[\text{heat}]{\text{cool}} Solid_2$$

(3) eutectoid reaction:

$$Solid_2 \xrightleftharpoons[\text{heat}]{\text{cool}} Solid_1 + Solid_3$$

(4) peritectoid reaction: $Solid_1 + Solid_3 \underset{heat}{\overset{cool}{\rightleftarrows}} Solid_2$

A peritectoid is a rare, reversed eutectoid. No peritectoid exists in this system.

(5) monotectoid reaction: $Liquid_2 \underset{heat}{\overset{cool}{\rightleftarrows}} Solid_1 + Liquid_3$

A monotectoid is a rare eutectic where the amount of one component in the eutectic composition is nil. No monotectoid exists in this system.

The eutectic at G (4.3 wt.% C) is a mixture of solid austenite (carbide dissolved in γ-iron, composition F) and solid carbide (cementite, composition H).

The peritectic at K is a mixture of solid δ-solution (carbide dissolved in δ-iron, composition J) and liquid solution (composition L).

The eutectoid at C (0.8 wt.% Carbon) yields pearlite, a lamillar structured solid mixture of α solid solution (composition B) and solid carbide (composition D).

The intersection at F indicates composition of the austenite component of eutectic G.

The false statement is (d) ●

MAT SCI 20

Steels can be strengthened by all of the following practices, EXCEPT:

(a)   annealing
(b)   quenching and tempering
(c)   work hardening
(d)   grain refinement
(e)   age or precipitation hardening

Annealing removes the strength increase caused by work hardening. Quenched and tempered steels rapidly get stronger as carbon content is increased to 0.6%, then more slowly above 0.6%. Alloys can increase strength by increasing hardenability (depth of hardening inward from the quenched surface). Work hardening results from cold drawing, rolling or extrusion performed below the recrystallization temperature. 600,000 psi tensile strength wire is produced by such work hardening of high carbon steel. Combinations of alloying elements can produce fine grained steels that are harder and tougher than course grained steels of the same composition.

The false statement is (a) ●

MAT SCI 21

If 1080 steel (0.80 wt.% C) is annealed by very slow cooling from $1000\,^{\circ}C$ ($1832\,^{\circ}F$) to ambient temperature, its microstructure will consist almost solely of:

(a)   austenite
(b)   bainite
(c)   cementite
(d)   martensite
(e)   pearlite

1080 steel in equilibrium at 1000°C has the austenite structure. During very slow annealing to ambient, diffusional transformation to pearlite occurs. On rapid quenching to ambient, a shear transformation yields metastable martensite. On tempering martensite at 200 - 400°C, fine grained bainite occurs enroute to course pearlite of the same α-ferrite plus carbide composition. Cementite is iron carbide (Fe_3C) containing 6.67 wt.% carbon.

Answer is (e) ●

MAT SCI 22

All of the following statements about the austenite-martensite-bainite transformations in steel are correct, EXCEPT:

(a) Martensite is fine grained α-ferrite, supersaturated with carbon, in a metastable body centered tetragonal structure. It forms by shear (slippage) during the rapid quench of face centered cubic austenite (δ -ferrite).

(b) Pearlite is a stable and course grained lamillar mixture consisting of body centered cubic α-ferrite plus carbide. It forms by eutectoidal transformation during the slow annealing of austenite. Most alloying elements in steel tend to retard this eutectoidal transformation.

(c) Martensite is strong and hard, but brittle. Tempering toughens it and reduces brittleness.

(d) Tempering of martensite is done by judicious reheating to produce, by diffusion, a fine-grained tough, strong microstructure.

(e) Bainite and tempered martensite have distinctively different microstructures.

〰〰〰〰〰

Bainite and tempered martensite have very closely related fine-grained microstructures. Pearlite, bainite and tempered martensite all consist of α-ferrite plus carbide. Bainite is usually formed from austenite by partial quench, to avoid pearlite formation, followed by isothermal transformation to fine-grained, tough, strong bainite. Differences between bainite and tempered martensite are subtle.

The false statement is (e) ●

MAT SCI 23

All of the following statements about <u>mechanical failure</u> are true, EXCEPT:

(a) Brittle fracture occurs with little plastic deformation and relatively small energy absorption.

(b) Ductile fracture is characterized by significant amounts of energy absorption and plastic deformation (evidenced by elongation and reduction in cross-sectional area).

(c) Ductile-brittle transition in failure mode occurs at reduced temperatures for most materials, because fracture strength remains constant with temperature while yield strength increases as temperature is reduced. At high temperatures yield strength is least; at low temperatures fracture strength is least.

(d) Fatigue failure due to cyclic stress is frequency dependent.

(e) Creep is time-dependent plastic deformation which accelerates at increased temperature. Stress rupture is the failure resulting from creep.

〰〰〰〰〰

All are correct except (d). Fatigue failure is not dependent on frequency at ambient temperature. It is very sensitive to surface imperfections from which cracks originate and propagate.

The false statement is (d) ●

MAT SCI 24

All of the following statements about rusting of iron are correct, EXCEPT:

(a) Contact with water, and oxygen are necessary for rusting to occur.

(b) Contact with a more electropositive metal reduces rusting.

(c) Halides aggravate rusting, a process which involves electrochemical oxidation-reduction reactions.

(d) Pitting occurs in oxygen-rich anodic areas, and the rust is deposited nearby.

(e) Strained regions are anodic relative to unstrained regions in the metal.

All except (d) are correct. Pitting occurs in oxygen depleted anodic areas. The localized rust deposit prevents further oxygen diffusion, further aggravating the situation. The cathodic area is not corroded. Cathodic protection makes iron the cathode; if corrosion occurs, it occurs at a sacrificial anode such as a magnesium rod or the zinc galvanizing.

The false statement is (d) ●

MAT SCI 25

Which of the following is not a method of non-destructive testing of steel castings and forgings?

(a) radiography
(b) magnetic particle
(c) ultrasonic
(d) liquid penetrant
(e) chemical analysis

All methods except (e) attempt to detect internal cracks and voids. Chemical analysis is destructive of the sample and does not detect structural flaws in the casting or forging.

The false answer is (e) ●

MAT SCI 26

Compressive strength of fully cured concrete is most directly related to:

(a) sand-gravel ratio
(b) fineness modulus
(c) aggregate gradation
(d) absolute volume of cement
(e) water-cement ratio

Sand and fine material (under 1/4 inch) are "fine aggregate", while gravel and crushed rock (to 3 inches) are "coarse aggregate".
Cement after hydration becomes the adhesive that binds a closely packed aggregate mixture, hence sand-gravel ratio must be optimized to obtain close packing for maximum strength. Fineness modulus is indicative of

the sieve analysis of the fine aggregate, and is used in calculating the ratio of course to fine aggregates used in the mix. Aggregate gradation affects slump or workability, with the larger sizes allowed in thicker sections.

Batch yields are calculated using absolute volumes of cement, water and aggregates. Usual specific gravities are 2.6-2.7 for sand and gravel, and 3.1 for cement.

<u>Water-cement ratio</u> is the most important factor affecting strength of the concrete; less water yields a higher strength, though less workable (lower slump) concrete.

Answer is (e) ●

MAT SCI 27

According to the ACI code, the modular ratio, n, of structural concrete with a 28-day ultimate compressive strength, f'_c, of 3000 is nearest to:

(a) 7
(b) 8
(c) 9
(d) 10
(e) 11

The modular ratio n is the ratio of elastic moduli of steel to concrete, E_s/E_c.

E_s = 29,000,000 psi, while E_c is calculated from concrete weight w and and strength f'_c using: $E_c = w^{1.5} \times 33(f'_c)^{0.5}$

The strength f'_c ranges from 2000 - 5000 psi, while concrete weight W varies with aggregate density. Lightweight aggregates (dry weights under 70 lb/ft^3) yield structural lightweight concrete of 90 - 115 lb/ft^3, while sand and gravel (about 100 lb/ft^3) yield structural concrete at about 145 lb/ft^3.

$$E_c = 145^{1.5}(33)(3000)^{0.5} = 3.2 \times 10^6 \text{ psi}$$

$$n = \frac{E_s}{E_c} = \frac{29 \times 10^6}{3.2 \times 10^6} = 9$$

Answer is (c) ●

MAT SCI 28

All of the following statements about air entrained concrete are correct, EXCEPT:

    (a) Air entrainment is recommended when concrete is exposed to severe frost action.

    (b) With air entrainment, the quantity of water to produce a given consistency (slump) is reduced.

    (c) With air entrainment, the quantity of water to produce a specified 28 day compressive strength is reduced.

    (d) Air entrainment reduces friction between solids during mixing.

    (e) Air entrainment reduces resistance to the freeze-thaw that occurs when salt is used to melt ice or snow.

The amount of entrained air varies inversely with maximum aggregate size, ranging from 0.2 to 3% without, and 3 to 8% with an added air entrainment agent. Voids reduce strength, and for a given water-cement ratio, air entrainment can reduce strength as much as 20%. Air entrainment does <u>increase</u> resistance to freeze-thaw conditions.

The false statement is (e) ●

MAT SCI 29

In the design of a reinforced concrete structure, tensile strength of the concrete is normally:

    (a) Assumed to be 1/10 of the 28-day compressive strength.

    (b) Determined by beam tests.

    (c) Neglected.

    (d) Assumed to be 200 psi.

    (e) Determined by split cylinder tests.

Hairline cracks can reduce tensile strength to zero. Reinforced concrete design by conventional cracked section analysis assumes <u>no tensile strength</u> in the concrete and takes all tensile forces in the steel. (Split cylinder tests do yield design data that are used to guard against diagonal tension failure resulting from shear stresses.) Prestressed concrete is reinforced concrete with imposed compressive stresses such that tensile stresses from service loading are opposed and the concrete remains in compression.

Answer is (c) ●

MAT SCI 30

All of the following groups of plastics are thermoplastic, EXCEPT:

    (a) Polyvinylchloride (PVC) and polyvinyl acetate

    (b) Polyethylene, polypropylene, and polystyrene

    (c) Tetrafluoroethylene (Teflon) and other fluorocarbons

    (d) Phenolics, melamine and epoxy

    (e) Acrylic (Lucite) and polyamide (nylon)

All are thermoplastic except group (d), which are thermosetting. Thermoplastic materials are completely polymerized, but soften for molding at elevated temperatures.

Thermosetting materials complete their polymerization during heating to give hard plastics.

All have different maximum service temperatures, above which decomposition begins. Most plastics become brittle at low temperatures; plasticizers are added to reduce low-temperature brittleness and improve ductility.

The false answer is (d) ●

6
FLUID
MECHANICS

FLUIDS 1

Kinematic viscosity may be expressed as:

(a) $\dfrac{ft^2}{sec}$

(b) $\dfrac{sec^2}{ft}$

(c) $\dfrac{slug\ sec^2}{ft}$

(d) $\dfrac{slugs}{sec}$

(e) $\dfrac{slug\ ft}{sec^2}$

Kinematic viscosity $\nu = \dfrac{absolute\ viscosity}{density} = \dfrac{\mu}{\rho}$

Units of viscosity: lb_f-sec/ft^2 or slug/ft-sec

Units of density: lb_f-sec^2/ft^4 or slugs/ft^3

In either units the dimensions of kinematic viscosity would be:

$$\dfrac{\dfrac{slug}{ft\text{-}sec}}{\dfrac{slug}{ft^3}} = \dfrac{ft^2}{sec} \quad\bullet\qquad \dfrac{lb_f\text{-}sec/ft^2}{lb_f\text{-}sec^2/ft^4} = \dfrac{ft^2}{sec} \quad\bullet$$

slug = $lb_f sec^2$/ft = lb_m/g_c Answer is (a)

FLUIDS 2

Absolute viscosity of a fluid varies with pressure and temperature and is defined as a function of

(a) Density and angular deformation rate.
(b) Density, shear stress and angular deformation rate.
(c) Density and shear stress.
(d) Shear stress and angular deformation rate.
(e) Angular deformation rate only.

<u>Absolute viscosity</u> is proportional to shear stress divided by angular deformation rate. Density is not involved in the definition.

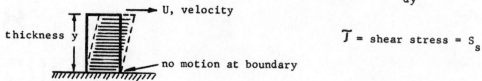

$$\text{rate of angular deformation} \cong \frac{dU}{dy}$$

$\mathcal{T}$ = shear stress = S_s

$$\text{absolute } \mu = \frac{S_s \text{ or } \mathcal{T}}{\frac{dU}{dy}} = \frac{lb_f/ft^2}{\frac{ft/sec}{ft}} = \frac{lb_f \text{ sec}}{ft^2} = \frac{slug}{ft \text{ sec}} \quad \text{in english units}$$

$$1 \; \frac{slug}{ft \text{ sec}} = 479. \text{ poise}, \quad \text{poise} = \frac{dyne \text{ sec}}{cm^2} = \frac{gm}{cm \text{ sec}}$$

For Newtonian fluids the relation between shear stress and angular deformation rate is linear; for non-Newtonian and thixotropic materials it is non-linear.

Answer is (d)●

FLUIDS 3

An open chamber rests on the ocean floor in 160. feet of sea water (sp. gr. = 1.03). What air pressure in psig must be maintained inside to exclude water?

    (a) 45.2
    (b) 60.9
    (c) 71.4
    (d) 93.2
    (e) 140.0

Internal pressure must equal external pressure.

External P = 160(62.4)(1.03) = 10,280 lb_f/ft^2

$$P = \frac{10,280}{144} = 71.4 \text{ psig} \; ●$$

Answer is (c)

FLUIDS 4

What is static gage pressure in psf in the air chamber of the container below?

    (a) -200. lb_f/ft^2
    (b)   -75. lb_f/ft^2
    (c)   0
    (d) +75. lb_f/ft^2
    (e) +200. lb_f/ft^2

Since the situation is static, gage pressure at the base is 5' water.
In the air chamber it is 5' water less 3' water less 4' oil.

$$P = 2' \text{ water} - 4' \text{ oil} = 2(62.4) - 4(62.4)(0.80)$$
$$= 124.8 - 199.7 = -74.9 \text{ lb}_f/\text{ft}^2 \bullet$$

Answer is (b)

FLUIDS 5

Pressure in lb_f/ft^2 at a depth of 300 feet in fresh water is nearest to:

(a)   5,360 psf
(b)   12,900 psf
(c)   18,700 psf
(d)   32,400 psf
(e)   56,500 psf

Pressure = weight of fluid above = $300(62.4) = 18,720 \text{ lb}_f/\text{ft}^2 \bullet$

Answer is (c)

FLUIDS 6

What head in feet of air, at ambient conditions of 14.7 psia and 68°F, is
equivalent to 2 psia?

(a)   146.
(b)   395.
(c)   771.
(d)   1560.
(e)   3840.

Density of air is calculated from ideal gas law using $R = 53.34 \dfrac{\text{ft lb}_f}{\text{lb}_m \text{ }°R}$

$$\rho = \frac{P}{RT} = \frac{(14.7 \text{ lb/in}^2)(144 \text{ in}^2/\text{ft}^2)}{53.34 \text{ ft lb}_f/(\text{lb}_m \text{ }°R)(528°R)} = 0.0752 \frac{\text{lb}_m}{\text{ft}^3}$$

2 psia = 2(144) = 288 lb/ft^2

$$h = \frac{288}{0.0752} = 3830 \text{ feet} \bullet$$

note that specific weight of air
$\gamma = \rho (g/g_c) = 0.0752 \text{ lb}_f/\text{ft}^3$

Alternately: 1000 feet of air at ambient conditions is approximately
equivalent to 0.5 psia.

Answer is (e)

FLUIDS 7

With a normal barometric pressure at sea level, atmospheric pressure at an
elevation of 4000 feet is nearest to:

(a)   26" Hg.
(b)   27" Hg.
(c)   28" Hg.
(d)   29" Hg.
(e)   30" Hg.

Normal barometric pressure at sea level is 29.92" Hg.
Near sea level this is reduced by about 1" Hg or 0.5 psia per 1000 feet.

The rate of pressure reduction diminishes with increased altitude since air
density decreases with altitude.

Answer is (a) $\bullet$

FLUIDS 8

The funnel shown is full of water. The volume of the upper part is 5.83 cu. ft. and of the lower part is 2.50 cu. ft. The force tending to push the plug out is:

      (a)   250 lbs.
      (b)   312 lbs.
      (c)   380 lbs.
      (d)   442 lbs.
      (e)   500 lbs.

$$\text{Force} = \gamma hA = 62.4 \ \frac{lb_f}{ft^3} \times 10 \text{ ft} \times 0.5 \text{ ft}^2 = 312 \ lb_f \ \bullet$$

Answer is (b)

FLUIDS 9

In the sketch shown below, a barrel whose empty weight is 56 lbs. contains an unknown quantity of water in which a container holding 200 lbs. of oil (S.G. = 0.95) floats, and a beaker containing 10 lbs. of mercury (S.G. = 13.0) which floats in the oil. Also, an iron rod is immersed in the water so that it displaces 1.0 cu. ft. of the liquid.

If the beaker, the oil container, and the barrel are 6 inches, 24 inches, and 34 inches in diameter, respectively, and the total weight indicated by the scale is 840 lbs, determine the height h of the water in the barrel.

Equivalent volume of water in barrel

$$= \frac{840 \text{ lb} - 56 \text{ lb}}{62.4 \text{ lb/cu ft}} = \frac{784}{62.4} = 12.55 \text{ cu ft}$$

Determine height h for an equivalent volume of 12.55 cu ft

$$V = \pi r^2 h \qquad h = \frac{V}{\pi r^2} = \frac{12.55}{\pi \left(\frac{17}{12}\right)^2} = \frac{12.55}{6.31} = 1.99 \text{ feet} \; \bullet$$

FLUIDS 10

An open topped cylindrical water tank has a horizontal circular base 10. feet in diameter. When filled to a height of 8 feet, the force in lbs. exerted on its base is nearest to:

(a) 3,900.
(b) 7,800.
(c) 10,000.
(d) 26,000.
(e) 39,000.

Pressure at tank base = 8' water = 8(62.4) = 499 lb_f/ft^2

Area of tank base = $\frac{\pi}{4}(10)^2 = 25\pi = 78.5 \text{ ft}^2$

Force on tank base = PA = 499(78.5) = 39,200 lb_f. $\bullet$

Answer is (e)

FLUIDS 11

A cubical tank with 4 foot sides is filled with water. The force in lbs. developed on one of the vertical sides is nearest to:

(a) 500.
(b) 1000.
(c) 2000.
(d) 3000.
(e) 4000.

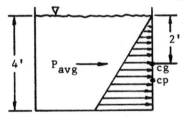

Average pressure exerted on one side is pressure existent at c.g. of the side times area of the side. Since pressure varies linearly with depth, the c.p. (center of pressure) is located below the c.g.

For other problems: $\quad y_{cp} = y_{cg} + \dfrac{I}{y_{cg}A}$

where y_{cg} = either vertical or slant distance in feet from free fluid surface to c.g. of area considered.

y_{cp} = distance to c.p., same vertical or slant distance basis, ft.

A = area considered, ft^2

I = area moment of inertia about centroid of area considered, ft^4

The c.p. is the location where a single concentrated force can be used to replace the distributed pressure forces for moment purposes. The pressure

existent at the area centroid times area equals force exerted normal to the
area acting through the c.p.

<u>For this problem</u>: $F = (P_{avg})(A) = 2(62.4)(4 \times 4) = 1997 \; lb_f$ ●

Answer is (c)

FLUIDS 12

Each connection between a 90 degree elbow and the 6" diameter pipeline
to which it is connected must resist what net tensile force in lbs. under
static, no-flow conditions if the line is pressurized to 100 psig?

6"D

A

6"D

B

(a)   710.
(b)   830.
(c)   2000.
(d)   2800.
(e)   4000.

Static pressure force at A: $F = (100 \; \frac{lb_f}{in^2}) \left[\frac{\pi}{4}(6)^2 \right] = 900\pi \; lb_f$ ↓

Static pressure force at B: $F = 900\pi \; lb_f$ ←

Connectors at each joint are under tensile force of 2827 lb_f ●

Answer is (d)

FLUIDS 13

A conical reducing section connects an existent 4" diameter pipeline with a
new 2"D line. At 100 psig static
pressure under no-flow conditions, what
tensile force in lbs. is exerted on the
connectors at joint A ?

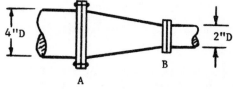

4"D

B

2"D

A

(a)   320.
(b)   470.
(c)   630.
(d)   940.
(e)   1260.

Static force at A = $(100 \; \frac{lb}{in^2}) \left[\frac{\pi}{4}(4)^2 \right] = 400\pi = 1257 \; lb_f$ tension on bolts at A   ●

Static force at B = $(100 \; \frac{lb}{in^2}) \left[\frac{\pi}{4}(2)^2 \right] = 100\pi = 314 \; lb_f$ tension on bolts at B

End restraint by the pipes opposes a net force of 1257-314=943 lb_f
to the right on the reducing section.

Answer is (e)

FLUIDS 14

A circular access port 2 feet in diameter seals an environmental test chamber that is pressurized to 15 psi above external pressure. What force in lbs. does the port exert upon its retaining structure?

(a) 1700.
(b) 2300.
(c) 3700.
(d) 5400.
(e) 6800.

Area of port = $\frac{\pi}{4} D^2 = \pi \ ft^2$

Pressure = $15 \ \frac{lb}{in^2} \times \frac{144 \ in^2}{ft^2} = 2160 \ lb_f/ft^2$

$F = PA = \pi (2160) = 6786 \ lb_f$ ●

Answer is (e)

FLUIDS 15

A vertical sliding gate 20' wide and 30' high is submerged in 40' of water, has a coefficient of friction μ between its guides and its edges of 0.20, and weighs 6 tons. What vertical force is required to just lift this gate?

(Assume friction is due to the normal force of the water pressure on the gate and neglect any buoyant force of the water.)

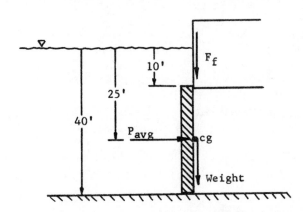

(1) Horizontal force against gate = (average pressure = pressure at area centroid) x area

c.g. is 25' below surface. $P = 25(62.4) = 1560 \ lb_f/ft^2$

$F = PA = 1560(20)(30) = 936,000. \ lb_f$

(2) Vertical frictional force $F_f = \mu N$ where μ = coeff. of friction = 0.2

N = force normal or perpendicular to gate = 936,000. lb_f

$F_f = (0.2)(936,000) = 187,200 \ lb_f$

(3) Total force to start gate upward = its weight + 187,200 lb_f.

$$= 12,000 + 187,200 = 199,200 \text{ lb} \bullet$$

FLUIDS 16

Commercial propane stored in a spherical steel tank at 100°F generates a gage pressure of 160 psi. If the tank is 4 feet in diameter and has walls 0.25 inch thick, what maximum tensile stress in psi is developed in the steel?

(a) 1,700.
(b) 2,900.
(c) 5,400.
(d) 7,700.
(e) 13,000.

A spherical tank, with no cylindrical section, is twice as strong as a cylindrical tank with hemispherical ends.

$$S_T = \frac{1}{2} \frac{Pr}{t} \quad \text{where} \quad P = \text{internal-external pressure difference}$$
$$r = \text{radius}$$
$$t = \text{wall thickness}$$
$$S_T = \text{tensile stress in the wall}$$

all in consistent units.

$$S_T = \frac{1}{2}(160 \ \frac{lb_f}{in^2})(\frac{24 \text{ in}}{0.25 \text{ in}}) = 7,700 \text{ psi} \bullet$$

Answer is (d)

FLUIDS 17

The steel cylindrical butane storage tank shown in the drawing has hemispherically domed ends. If butane has a vapor pressure of 45. psia at 100°F, and the tank walls are 0.25 inches thick, the maximum tensile stress in psi developed in the tank is nearest to:

(a) 4,400.
(b) 7,300.
(c) 9,700.
(d) 12,000.
(e) 17,000.

Mode of failure will be a longitudinal rip in the cylindrical section.

Maximum stress will be tensile hoop stress given by $S_T = \frac{Pr}{t}$

P is the internal-external pressure differential = 45 - 14.7 = 30.3 psig

$$S_T = 30.3 \text{ psi}(\frac{36}{0.25}) = 4,360 \text{ psi} \bullet$$

Answer is (a)

FLUIDS 18

A gas bubble rising from the ocean floor is 1 inch in diameter at a depth of 50 feet. Given that sp. gr. of seawater is 1.03, the buoyant force in lbs being exerted on the bubble at this instant is nearest to:

(a) 0.014
(b) 0.020
(c) 0.076
(d) 0.14
(e) 0.20

Volume of bubble = volume of displaced seawater

$$= \frac{4}{3}\pi r^3 = \frac{4}{3}\pi\left(\frac{1}{2}\right)^3 = \frac{\pi}{6} = 0.524 \text{ in}^3$$

density of seawater = $62.4(1.03)\frac{lb_m}{ft^3} \times \frac{ft^3}{1728 \text{ in}^3} = 0.0372 \frac{lb_m}{in^3}$. Specific weight

of seawater is $0.0372 \text{ lb}_f/\text{ in}^3$.
Ignore sp. wt. of bubble calculated at water temperature and an absolute
pressure of atmospheric + $50(1.03)(62.4) \text{ lb}_f/\text{ft}^2$.

Buoyant force = $0.524 \text{ in}^3(0.0372 \frac{lb_f}{in^3}) = 0.0195 \text{ lb}_f$ ●

Answer is (b)

FLUIDS 19

Draft, in inches of water differential pressure, is generated at the base of
a 100 foot stack filled with 500°F gases (assume same molecular weight as air)
due to differential specific weight.

(Under ambient conditions of 14.7 psia and 68°F, air sp. wt. is $0.075 \text{ lb}_f/\text{ft}^3$.)

The draft is nearest to: (a) 0.30" water
                            (b) 0.65" water
                            (c) 1.00" water
                            (d) 1.50" water
                            (e) 3.00" water

Sp. wt. of stack gases = $0.075 \frac{528°R}{960°R} = 0.0413 \text{ lb}_f/\text{ft}^3$

external P at 100' elev.

$\Delta P = 100(0.075 - 0.041) = 3.4 \text{ lb}_f/\text{ft}^2$

$1"\text{ water} = \frac{62.4}{12} = 5.2 \text{ lb}_f/\text{ft}^2$

$\therefore \Delta P = \frac{3.4}{5.2} = 0.65"\text{ water}$ ●

Answer is (b)

FLUIDS 20

Ice in an iceberg has a sp. gr. of 0.922. When floating in seawater
(sp. gr. 1.03), its exposed volume % is nearest to:

      (a) 5.6
      (b) 7.4
      (c) 8.9
      (d) 10.5
      (e) 11.8

Buoyant force equals weight of fluid displaced. At equilibrium or floating,
weight downward equals buoyancy.

Let V_1 = total volume of iceberg in ft^3

Its weight is $V_1(62.4)(0.922) = 57.5(V_1)$ lb$_f$

Let V_2 = immersed volume of iceberg = volume of seawater displaced.

Weight of seawater displaced = $V_2(62.4)(1.03) = 64.2(V_2)$ lb$_f$

$$\frac{V_2}{V_1} = \frac{57.5}{64.2} = 0.895 = \text{volume fraction of iceberg immersed}$$

∴ volume fraction exposed = 1. - 0.895 = 0.105 = 10.5% ●

Answer is (d)

FLUIDS 21

A cylinder of cork is floating upright in a container partially filled with water. A vacuum is applied to the container such that the air within the vessel is partially removed. The cork will

      (a) rise somewhat in the water
      (b) sink somewhat in the water
      (c) remain stationary
      (d) turn over on its side
      (e) sink to the bottom of the container

Archimedes' Principle applies equally well to gases. Thus a body located in any fluid, whether liquid or gaseous, is buoyed up by a force equal to the weight of the fluid displaced. A balloon filled with a lighter than air gas readily demonstrates the buoyant force.

Thus the weight of the cork is equal to the weight of water displaced plus the weight of air displaced. When the air within the vessel is removed, the cork is no longer provided a buoyant force equal to the weight of air displaced. For equilibrium, the cork will sink somewhat in the water. ●

Answer is (b)

FLUIDS 22

A floating cylinder 8 cm in diameter and weighing 950 grams is placed in a cylindrical container 20 cm in diameter partially full of water. The increase in the depth of water in the container due to placing the float in it is:

      (a) 10 cm
      (b) 5 cm
      (c) 3 cm
      (d) 2 cm
      (e) 1 cm

The 950 gram cylinder will displace 950 grams of water. Since 1 cc of water weighs 1 gram, the cylinder will displace 950 cc of water. The change in total volume beneath the water surface dV equals the area of the cylindrical container A times the change in water level dh, or dV = A dh.

The depth of the water will increase

$$dh = \frac{dV}{A} = \frac{950 \text{ cm}^3}{\frac{\pi}{4}(20)^2} = 3.02 \text{ cm} ●$$

Answer is (c)

FLUIDS 23

A block of wood floats in water with 6 inches projecting above the water surface. If the same block were placed in alcohol of specific gravity 0.82, the block would project 4 inches above the surface of the alcohol. Determine the specific gravity of the wood block.

Let x = height of wood block

W = width of wood block

L = length of wood block

γ = specific weight of water, lb_f/ft^3

The weight of the block is equal to the weight of the liquid displaced.

Weight of the block in water = $(x - 6)WL\gamma$ 1.0

Weight of the block in alcohol = $(x - 4)WL\gamma$ 0.82

Since the weight of the block is constant:

$(x - 6)WL\gamma = 0.82(x - 4)WL\gamma$

$$x - 6 = 0.82x - 3.28 \qquad x = \frac{2.72}{0.18} = 15.1 \text{ inches}$$

The Specific Gravity of the wood block:

$$= \frac{\text{Volume of water displaced}}{\text{total volume}} = \frac{(x - 6)WL}{xWL} = \frac{15.1 - 6}{15.1} = 0.603 \; \bullet$$

FLUIDS 24

Average velocity in a full pipe of incompressible fluid at Section 1 of the drawing is 10 ft/sec. After passing through a conical section that reduces the stream's cross-sectional area at Section 2 to one fourth of its previous value, the velocity at Section 2, in ft/sec, is:

(a) 2.5
(b) 5.
(c) 10.
(d) 40.
(e) 80.

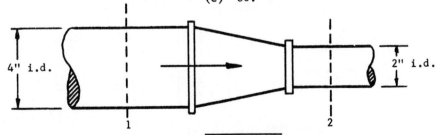

Continuity requires that:

$$q = A_1 v_1 = A_2 v_2 = A_1 (10) = \frac{A_1}{4} v_2 \qquad v_2 = 40. \text{ ft/sec} \; \bullet$$

Answer is (d)

FLUIDS 25

Refer to the drawing of the previous question.
If the static pressure at A is 100 psig and the 4" diameter pipe is full of
water in turbulent flow at an average velocity of 30. ft/sec at A, the mass
flow rate in lb/sec at B is nearest to:

      (a)   3.7
      (b) 48.
      (c) 160.
      (d) 270.
      (e) 520.

Basis of calculation: 1 second

Continuity requires that mass flow rate be the same at all sections in steady
flow.

Calculate $\dot{m}$ at Section A where velocity is given, using $\dot{m} = \rho Av$

cross-sectional area $A = \dfrac{\pi}{4}\left(\dfrac{4}{12}\right)^2 = 0.087$ ft^2

$$\dot{m} = \left(62.4 \ \frac{lb_m}{ft^3}\right)(0.087 \ ft^2)\left(30 \ \frac{ft}{sec}\right) = 163 \ lb_m/sec \ \bullet$$

Answer is nearest (c)

FLUIDS 26

Air flows in a long length of 1" diameter pipe. At one end the pressure is
30 psia, the temperature is 300°F and the velocity is 30 ft/sec.

At the other end the pressure has been reduced by friction and heat loss to
20 psia. The mass flow rate in lb/sec at any section along the pipe is
nearest to:

      (a) 0.02
      (b) 0.11
      (c) 0.37
      (d) 1.5
      (e) 3.9

Mass flow rate $\dot{m} = q\rho = \rho Av$ Basis of calculation: 1 second

Density of air at 30 psia and 300°F (760°R) is obtained from ideal gas
law:

$$Pv = RT = \frac{P}{\rho} \qquad \text{Use } R = 53.34 \ ft \ lb_f/(lb_m \ {}^\circ R) \text{ for air}$$

$$\rho = \frac{P}{RT} = 30(144)\frac{lb_f}{ft^2} \ \frac{lb_m \ {}^\circ R}{53.34 \ ft \ lb_f (760 {}^\circ R)} = 0.107 \ \frac{lb_m}{ft^3}$$

$$A = \text{cross-sectional area} = \frac{\pi}{4}D^2 = 0.785\left(\frac{1}{12}\right)^2 = 0.00545 \ ft^2$$

$$\dot{m} = \rho Av = \left(0.107 \ \frac{lb_m}{ft^3}\right)\left(0.00545 \ ft^2\right)\left(30 \ \frac{ft}{sec}\right) = 0.0175 \ \frac{lb_m}{sec} \ \bullet$$

Answer is nearest (a)

FLUIDS 27

Water flows through a long $\frac{1''}{2}$ i.d. hose at 3 gallons per minute.

Water velocity in ft/sec is nearest to:

(a) 1.
(b) 5.
(c) 10.
(d) 20.
(e) 50.

Basis of calculation: 1 second

$$q = 3 \frac{gal}{min} \times \frac{min}{60 \ sec} \times \frac{ft^3}{7.48 \ gal} = 0.00668 \ cfs$$

$$A = \text{cross-sectional area} = \frac{\pi}{4} D^2 = 0.785 \left[\frac{0.5}{12}\right]^2 = 0.00136 \ ft^2$$

$$q = Av \qquad v = \frac{0.00668}{0.00136} = 4.9 \ ft/sec \ \bullet$$

Answer is nearest (b)

FLUIDS 28

An incompressible fluid gasoline (ρ = 50 lb/ft^3) enters and leaves a pump system with the following energy in ft lb$_f$ per lb$_m$ of fluid:

	Entering	Leaving
Potential energy Z ft above datum	5	15
$\frac{v^2}{2g_c}$ Kinetic energy	5	10
pv Flow energy	30	150
Total energy	40	175

The pressure increase in psi between entering and leaving streams is nearest to:

(a) 42.
(b) 84.
(c) 108.
(d) 120.
(e) 135.

Pressure (flow) energy change is 150 - 30 = 120 ft lb$_f$/lb$_m$

$$\Delta(Pv) = \frac{\Delta P}{\rho} = 120 \qquad \Delta P = 120 \frac{ft \ lb_f}{lb_m} \ 50 \frac{lb_m}{ft^3} = 6000 \frac{lb_f}{ft^2}$$

$$\Delta P = 6000 \frac{lb_f}{ft^2} \times \frac{ft^2}{144 \ in^2} = 41.7 \ psi \ \bullet$$

Answer is (a)

FLUIDS 29

Use the data of the previous question. If the volume flow rate of the gasoline ($\rho = 50$ lb/ft^3) is 15 gallons per minute, theoretical pumping power in HP is nearest to:

(a) 0.4
(b) 0.8
(c) 1.2
(d) 3.4
(e) 6.9

Basis of calculation: 1 minute

$$\text{mass flow rate} = 15 \frac{\text{gal}}{\text{min}} \times \frac{\text{ft}^3}{7.48 \text{ gal}} \times \frac{50 \text{ lb}_m}{\text{ft}^3} = 100. \frac{\text{lb}_m}{\text{min}}$$

Ignoring head loss h_L due to friction, the required

energy input is 175 - 40 = 135. ft lb$_f$/lb$_m$

$$HP = \frac{100 \text{ lb}_m}{\text{min}} \times \frac{135 \text{ ft lb}_f}{\text{lb}_m} \times \frac{1}{33,000 \text{ ft lb}_f/(\text{HP min})} = 0.41 \text{ HP} \bullet$$

Answer is (a)

FLUIDS 30

Water flowing in a pipe enters a horizontal venturi tube whose throat area at B is 1/4 that of the original and final cross-sections at A and C.

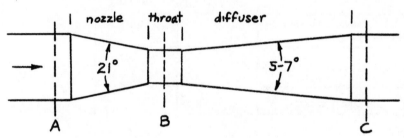

Continuity and energy conservation demand that which one of the following be true?

(a) the pressure at B is increased.
(b) the velocity at B is decreased.
(c) the potential energy at C is decreased.
(d) the flow energy at B is decreased.
(e) kinetic energy at B is reduced.

In a venturi throat the increased velocity, required by continuity, results in a KE increase that occurs at the expense of pressure (flow) energy. $\bullet$

Since the system is horizontal, no change in potential energy has occurred. At B the pressure (flow energy) decreases and velocity (KE) increases.

For a well designed venturi, conditions existent at A are essentially restored.

Answer is (d)

FLUIDS 31

Given the following energy data in ft lb_f/lb_m existent at two sections across a pipe transporting water in steady flow:

	Section A	Section B
Potential energy	65	135
Kinetic energy	50	50
Flow energy	335	245
TOTAL	450	430

What frictional head loss in feet has occurred?

(a)   0.
(b)   20.
(c)   70.
(d)   90.
(e)   160.

By energy balance on the basis of 1 lb_m of fluid flowing:

Total energy in = Total energy out + Energy losses - Energy inputs

$$450 = 430 + h_L - 0$$

Head loss, $h_L = +20$ ft lb_f/lb_m, or 20 feet. ●

Answer is (b)

FLUIDS 32

Theoretical horsepower required to pump water at 100 gallons per minute from a large reservoir to the surface of another large reservoir 400 feet higher is nearest to:

(a)   6. HP
(b)   10. HP
(c)   18. HP
(d)   35. HP
(e)   46. HP

Ignoring frictional losses, pump inefficiency, and any changes in KE or pressure, pumping power is equal to the increase in potential energy given the water.

Z or h = 400 ft lb_f/lb_m potential energy increase   per lb_m.

$\dot{m}$ = mass flow rate = $100 \frac{gal}{min} \times 8.33 \frac{lb_m}{gal} = 833$ lb_m/min

Basis of calculation: 1 minute

$$HP = \frac{400 \times 833 \text{ ft lb/min}}{33,000 \text{ ft } lb_f/(HP \text{ min})} = \frac{333,200}{33,000} = 10.1 \text{ HP} ●$$

Answer is (b)

FLUIDS 33

A horizontal stream of water with a cross sectional area of 0.1 ft^2 and a velocity of 40. ft/sec has kinetic energy in ft lb$_f$/sec nearest to:

    (a)   2,500.
    (b)   6,200.
    (c) 11,700.
    (d) 56,000.
    (e) 132,000.

Volume flow rate, q = Av = 0.1(40) = 4.0 cfs

Mass flow rate, $\dot{m} = q\rho = 4.0(62.4 \, \frac{lb_m}{ft^3}) = 249.6 \, lb_m/sec.$

Kinetic energy $= \frac{v^2}{2g}$ ft lb$_f$/lb$_m$

Per second: KE $= \left(249.6 \, \frac{lb_m}{sec}\right)\left(\frac{1600}{64.4}\right) = 6200.$ ft lb$_f$/sec. ●

Answer is (b)

FLUIDS 34

The theoretical velocity generated by a 10 foot hydraulic head is:

    (a) 12.2 ft/sec
    (b) 17.9 ft/sec
    (c) 25.4 ft/sec
    (d) 29.2 ft/sec
    (e) 35.8 ft/sec

$V = (2gh)^{1/2} = (2 \times 32.2 \times 10)^{1/2} = (644)^{1/2} = 25.4$ ft/sec. ●

Answer is (c)

FLUIDS 35

What is the static head corresponding to a flow velocity of 10 ft/sec?

    (a) 1.55 ft
    (b) 1.75 ft
    (c) 2.05 ft
    (d) 2.25 ft
    (e) 2.50 ft

$h = \frac{v^2}{2g} = \frac{10^2}{2(32.2)} = 1.55$ ft. ●

Answer is (a)

FLUIDS 36

Total energy of a compressible or incompressible fluid flowing across any section in a pipeline is a function of:

    (a) pressure and velocity.

    (b) pressure, density and velocity.

    (c) pressure, density, viscosity and velocity.

    (d) pressure, density, velocity, and height above datum.

    (e) flow energy, kinetic energy, height above datum, and internal energy.

Total energy for a 1st law of thermodynamics energy balance contains all relevant energy terms:

Potential energy, based on height above datum.

Kinetic energy (dynamic or velocity energy), based on velocity.

Flow energy, or pressure energy, based on pressure and density or specific volume.

Internal energy, based on temperature and heat capacity.

The Bernoulli equation for incompressible fluids ignores thermal energy (internal energy), although widely used.

Answer is (e) ●

FLUIDS 37

The locus of the elevations to which water will rise in a piezometer tube is termed:

      (a)   stagnation pressure
      (b)   the energy gradient
      (c)   the hydraulic gradient
      (d)   friction head
      (e)   critical depth

Piezometer tube is a static pressure indicating tube, equivalent to a static pressure gage.

Stagnation pressure is an increased pressure developed on impact with a pitot tube as a result of localized kinetic energy reduction to zero.

Hydraulic gradient or hydraulic grade line is flow energy or pressure head in ft lb_f/lb_m plotted vertically above the pipe centerline along the pipe length.

Energy gradient or energy grade line is total energy (flow energy or pressure head, plus kinetic energy or dynamic head, plus potential energy or height above datum) in ft lb_f/lb_m plotted vertically above the elevation datum along the pipe length.

Friction head is head loss h_L in ft lb_f/lb_m under flow conditions.

Critical depth above channel floor in open channels is depth for minimum potential + kinetic energy for the volume flow rate the channel delivers. Tranquil flow (at low KE and high PE) exists above critical depth, and rapid flow (at high KE and low PE) exists below critical depth.

Answer is (c) ●

FLUIDS 38

A water tank consists of a right circular cone that has its central axis vertical and its vertex at the bottom.

The radius at the top is 10 feet, the height is 50 feet, and the water is 15 feet deep.

Determine the work required to pump all the water to the level of the top of the cone.

$$W = \rho \int h dV$$

For the cone the differential element of volume is

$$dV = \pi r^2 dy$$

For water $\rho = 62.4$ lbs/ft^3

The radius r is changing by the relation

$$r = \frac{10}{50} y = 0.2y$$

Therefore

$$dV = \pi (0.2y)^2 dy$$

The distance through which the element of volume must be moved $h = (50 - y)$

$$W = 62.4 \int_0^{15} \pi (0.2y)^2 (50 - y) dy$$

$$= 62.4 \, \pi \, (0.04) \int_0^{15} 50y^2 - y^3 \, dy$$

$$= 7.83 \left[\frac{50y^3}{3} - \frac{y^4}{4} \right]_0^{15}$$

$$= 7.83 \left[16.67(y^3) - 0.25(y^4) \right]_0^{15}$$

$$= 342 \times 10^3 \text{ ft-lbs.} \bullet$$

FLUIDS 39

A stream of fluid with a mass flow rate of 2 slugs/sec and a velocity of 20 ft/sec to the right has its direction reversed 180° in a U fitting. The net dynamic force in lb$_f$ exerted by the fluid on the fitting is nearest to:

      (a)   40.
      (b)   80.
      (c)   514.
      (d) 1288.
      (e) 2576.

Apply impulse-momentum $Fdt = d(mv)$ in integrated form on the basis of 1 second:

      F lbs (1 sec) = (mass flow rate, slugs/sec)($v_2 - v_1$, ft/sec)

      $F = 2(-20 -20) = -80$ lb$_f$

Since original velocity was taken as +20 in a positive direction, final reversed velocity is -20. Force from impulse-momentum is the force on the fluid necessary to accomplish the velocity change. By reaction, the fluid exerts an equal and opposite force $+ 80$ lb$_f$ to the right on the fitting. $\bullet$

            Answer is (b)

FLUIDS 40

The thrust in lb_f generated by an aircraft jet engine on takeoff per 1 lb_m/sec
of exhaust products, whose velocity has been increased from essentially 0 to
500 ft/sec, is nearest to which of the following:

(a)   15.
(b)   130.
(c)   360.
(d)   710.
(e) 2200.

The combustion products are predominantly nitrogen from the intake air
(approx. 15 lb air/lb fuel). Use impulse-momentum:

$$F\ dt = d(mv)$$

Change in momentum = momentum of outlet stream less momentum of air and fuel
        inlet streams.

$$\Delta(mv) = m_{products}(500) - m_{air}(0) - m_{fuel}(0)$$

Evaluate at 1 second:

$$F = \dot{m}(v_2 - 0) = \left(\frac{1}{32.2}\ \frac{slug}{sec}\right)\left(500\ \frac{ft}{sec}\right) = 15.5\ lb_f\ \bullet$$

Answer is (a)

FLUIDS 41

A jet of water issues vertically upward from a nozzle with a velocity V ft/sec
and a flow rate Q ft^3/sec. Vertically above the nozzle at a distance h feet
a horizontal plate is placed as shown.

If the density of the water is w lb/ft^3, what reaction force F is required to
keep the plate stationary? Neglect the weight of the plate.

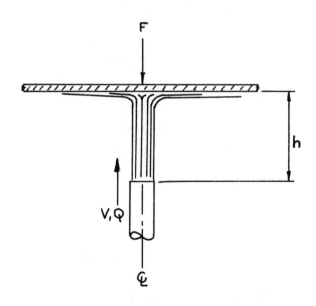

The water squirting upward loses velocity. Use an energy balance to get its velocity just prior to impinging on the plate.

$$\frac{v^2}{2g} - h = \frac{V_2^2}{2g} \qquad \text{or} \quad V_2 = \sqrt{v^2 - 2gh}$$

Then compute the force F by impulse-momentum for an elapsed time of 1 sec when Q ft^3 flows:

$$F \cdot dt = d(mV) = mdV + V \, dm$$

$$F \; lb_f(1 \; sec) = (Q \; ft^3)(\frac{w}{g} \frac{lb_f}{ft^3} \cdot \frac{sec^2}{ft})(V_2 - 0 \; \frac{ft}{sec})$$

$$F = Q \frac{w}{g} \sqrt{v^2 - 2gh} \quad lbs \quad \bullet$$

FLUIDS 42

Which of the following statements most nearly approximates conditions in turbulent flow?

(a) Fluid particles move along smooth, straight paths.
(b) Energy loss varies linearly with velocity.
(c) Energy loss varies as the square of the velocity.
(d) Newton's law of viscosity governs the flow.
(e) The Reynolds number is always less than 2000.

Laminar (streamline, viscous) flow is compared with turbulent flow in the table below.

	Laminar Flow	Turbulent Flow
Motion of fluid particles	Parallel to stream velocity. Paths of particles do not cross.	Particle paths cross and move in all directions.
Energy loss, h_f	$h_f = f\left(\frac{L}{D}\right)\left(\frac{v^2}{2g}\right)$ f is independent of surface roughness and decreases with Re. $f = \frac{64}{Re}$	$h_f = f\left(\frac{L}{D}\right)\left(\frac{v^2}{2g}\right)$ f varies with surface roughness, decreases with Re to constant value. See Moody diagram.
Velocity distribution in pipe	Average is 1/2 of max at centerline. Parabolic distribution. Zero at wall.	Essentially same throughout, except for thin boundary layer at wall. Follows 1/7 power law.
Reynolds number $Re = \frac{DV\rho}{\mu}$	Less than 2000	Greater than 2000

Newtons law of viscosity defines μ on the basis of shear stress and rate of angular deformation. Reynolds number involves μ as one contributing parameter. Very viscous liquids are usually in laminar flow.

On basis of above data, select (c). Do not confuse energy loss, h_f, with friction factor, f.

Answer is (c) ⬤

FLUIDS 43

For turbulent flow of a fluid in a pipe, all of the following are true, <u>except</u>:
- (a) The average velocity will be nearly the same as at the pipe center.
- (b) Energy lost due to turbulence and friction varies with kinetic energy.
- (c) Pipe roughness affects the friction factor.
- (d) Reynolds number will be less than 2000.
- (e) Flow of gases and low viscosity liquids is usually turbulent.

In turbulent flow the Reynolds number is <u>greater</u> than 2000. ●

Answer is (d)

FLUIDS 44

In fluid flow, if the fluid travels parallel to the adjacent layers and the paths of individual particles do not cross, the flow is said to be:
- (a) laminar
- (b) turbulent
- (c) critical
- (d) dynamic
- (e) uniform

Answer is (a) ●

FLUIDS 45

Flow nets are drawn to show all of the following except:
- (a) Streamlines spaced to bound streamtubes of equal volume flow rate, q.
- (b) Paths of particles in 2-and 3-dimensional flow.
- (c) Equipotential lines perpendicular to streamlines.
- (d) Equipotential lines spaced equal to distance between the streamlines they cross.
- (e) Flows and pressure drops in distribution systems.

All are true <u>except</u> (e). Pipe networks (containing parallel pipes, different size pipes in series, and branch pipes) are analyzed analogously to electrical networks.

Typical streamlines and equipotential lines, existent in 2-dimensional flow in the bend of a passage, are shown in the drawing.

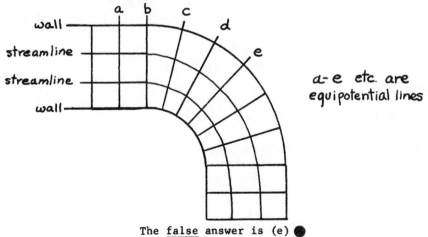

The <u>false</u> answer is (e) ●

FLUIDS 46

Which of the following constitutes a group of parameters with the dimensions of power?

(a) ρAV

(b) PAV

(c) $\dfrac{DV\rho}{\mu}$

(d) $\dfrac{\rho V^2}{P}$

(e) $\dfrac{V^2}{Dg}$

~~~~~~~~~~~~~~

(a) is mass flow rate, m in slugs/sec or lb_m/sec.

(b) Power has dimensions of ft lb_f/sec. PAV has these dimensions.

(c) is Reynolds number, Re, dimensionless, ratio of inertial force to viscous force.

(d) is Euler number, Eu, dimensionless, ratio of inertial force to pressure force.

(e) is Froude number, Fr, dimensionless, ratio of inertial force to gravity force.

Parameters involved are:

A, cross sectional area, ft^2

D, diameter, ft

P, pressure, lb_f/ft^2

V, velocity, ft/sec

$slug = lb_f\ sec^2/ft = \dfrac{lb_m}{g_c}$

μ = absolute viscosity, $lb_f\ sec/ft^2$
   or slug/(ft sec)

ρ = density, $lb_f\ sec^2/ft^4$ or lb_m/ft^3
   or $slug/ft^3$

g = acceleration due to gravity
   32.17 ft/sec^2

g_c = conversion factor = 32.17 $lb_m\ ft/(lb_f\ sec^2)$

Answer is (b) ●

FLUIDS 47

At or below critical velocity in small pipes or at very low velocities, the loss of head due to friction

(a) varies directly as velocity.
(b) can be ignored.
(c) is infinitely large.
(d) varies as the velocity squared.
(e) equals the velocity head.

~~~~~~~~~~~~~~

Answer is (a) ●

FLUIDS 48

Given the Moody diagram on the next page which is a log-log plot of friction factor vs. Reynolds number.

Which of the lines A-E represents the friction factor to use for a smooth pipe of low roughness ratio (ϵ/D) in streamline flow?

(a) A
(b) B
(c) C
(d) D
(e) E

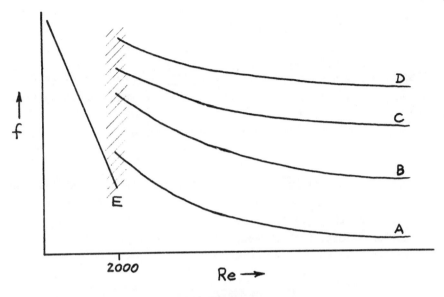

Line E applies to all roughness ratios in streamline flow (Re < 2000) because stagnant fluid layers at the wall make friction factor independent of roughness ratio; $f = \frac{64}{Re}$. In turbulent flow (Re > 2000) increasing roughness is represented by A for smooth pipe, to D for very rough pipe. In turbulent flow, where only the thinnest boundary layer exists, friction factor is very dependent on surface roughness.

Answer is (e) ●

FLUIDS 49

At constant pressure P all of the following statements about flow rate, q in turbulent flow in a pipe are true, <u>except</u>:

    (a) q varies approximately as (diameter, D)^2.

    (b) q decreases as pipe surface roughness ratio, $\frac{\epsilon}{D}$ increases.

    (c) q decreases as pipe length, L increases.

    (d) q decreases inversely as (fluid density, ρ)$^{1/2}$

    (e) q varies as (velocity, V)^2

At steady state, an energy balance equates flow (pressure) energy with frictional loss + KE. $\frac{P}{\rho} = h_L + \frac{V^2}{2g_c}$.

The Darcy-Weisbach equation quantifies loss: $h_L = f\left(\frac{L}{D}\right)\frac{V^2}{2g_c}$. The friction factor f is a function of Reynolds no., Re, and surface roughness, $\left(\frac{\epsilon}{D}\right)$ as shown on the Moody diagram of the previous question. In turbulent flow, f usually ranges from 0.01 to 0.06 and becomes constant at high Re. Since q = AV, an overall expression for flow can be assembled for circular pipes in terms of the parameters D, P, ρ, L and f.   Then f can be determined from Re $= \dfrac{DV\rho}{\mu}$ and $\dfrac{\epsilon}{D}$.

$$q = \frac{\pi}{4} D^2 \left[\frac{2g_c\, P}{\rho \left[1 + f\left(\frac{L}{D}\right)\right]} \right]^{1/2}$$

All of the problem statements are correct, except (e), for q = AV shows that q varies directly as velocity. Note that the derived equation shows that q varies as $P^{1/2}$

The <u>false</u> answer is (e) ●

FLUIDS 50

A 24-inch water pipe carries 15 cfs. At point A the elevation is 150 feet and the pressure is 30 psig. At point B, 4000 feet downstream from A, the elevation is 130 feet and the pressure is 35 psig.

Determine   (a)   head loss, h_L, between A and B.

(b)   HP loss between A and B.

(c)   friction factor, f.

〰〰〰〰〰

(1) Use energy balance to determine h_L. Basis: 1 lb_m of fluid.

PE + KE + Pressure (flow) energy = PE + KE + Pressure (flow) energy + h_L

$$150 + \frac{v^2}{2g_c} + \frac{30(144)}{62.4} = 130 + \frac{v^2}{2g_c} + \frac{35(144)}{62.4} + h_L$$

Since the pipe diameter is unchanged, continuity requires that V be the same at both points. KE can be deleted from both sides of the equation.

$$150 + 69.2 = 130 + 80.8 + h_L$$

$$h_L = 8.40 \quad \text{ft lb}_f/\text{lb}_m$$

(2) Calculate energy loss in HP. Basis: 1 second

mass flow rate, $\dot{m} = q\rho = \left(15\ \frac{\text{ft}^3}{\text{sec}}\right)\left(62.4\ \frac{\text{lb}_m}{\text{ft}^3}\right) = 936.\ \text{lb}_m/\text{sec}$

energy loss $= \left[936\ \frac{\text{lb}_m}{\text{sec}}\right]\left[8.40\ \frac{\text{ft lb}_f}{\text{lb}_m}\right] = 7862.\ \frac{\text{ft lb}_f}{\text{sec}}$

HP loss $= \dfrac{7862\ \text{ft lb/sec}}{550\ \text{ft lb/(HP sec)}} = 14.3$ HP

(3) Calculate velocity, V, and substitute in Darcy-Weisbach equation.

q = AV $A = \frac{\pi}{4}(2)^2 = \pi\ \text{ft}^2$ $V = \dfrac{15}{\pi} = 4.78$ ft/sec

$$h_L = f\left(\frac{L}{D}\right)\frac{v^2}{2g_c} \qquad 8.40 = f\left[\frac{4000}{2}\right]\frac{22.85}{2(32.17)}$$

$$f = 0.0118 \ ●$$

FLUIDS 51

Entrance losses between tank and pipe, or losses through elbows, fittings and valves are generally expressed as functions of

      (a) kinetic energy
      (b) pipe diameter
      (c) friction factor
      (d) volume flow rate
      (e) all of the above

Typical head losses for the above items are expressed as an empirical average constant, K, times kinetic energy, $\frac{v^2}{2g_c}$.$\qquad h_L = K \; \frac{v^2}{2g_c} \; = ft \; lb_f/lb_m.$

Values of K vary widely, but the list below is typical for minor losses:

	K
Entrance loss - flush with wall	0.50
- projecting inside tank	1.0
- smoothy contoured	0.05
Exit loss - entering tank	1.0
90° Elbow	0.90
Gate valve - fully open	0.20
Globe valve - fully open	10.0

Answer is (a) ●

FLUIDS 52

For the diagram shown, compute the velocity of the water flowing in the 1000' branch of the 6" diameter pipe.

Assume the friction factors in the two 6" round pipes are the same and that the incidental losses are equal in the two branches.

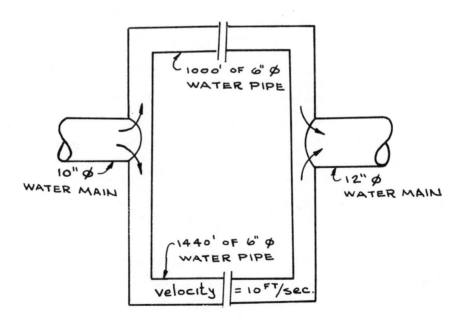

There is a drop in the energy line from the 10" main to the 12" main. This head loss must be equal in both 6" branches.

$$h_{L_{1000'}} + \text{incidental losses}_{1000'} = h_{L_{1440'}} + \text{incidental losses}_{1440'}$$

The Darcy equation is

$$h_L = f \frac{L}{d} \frac{v^2}{2g}$$

where h_L = head loss in feet

f = friction factor
L = length of pipeline in ft.
d = diameter of pipe in ft.
g = 32.2 ft/sec^2

Thus in this situation:

$$f \frac{1000'}{0.5'} \frac{v^2_{1000}}{2g} = f \frac{1440'}{0.5'} \frac{v^2_{1440}}{2g}$$

which reduces to

$$1000v^2_{1000} = 10^2(1440)$$

$$v_{1000'} = \sqrt{\frac{100(1440)}{1000}} = 12 \text{ ft/sec} \;\bullet$$

FLUIDS 53

The vena contracta of a sharp edged hydraulic orifice usually occurs:

   (a) At the geometric center of the orifice.
   (b) At a distance of about 10% of the orifice diameter upstream from the plane of the orifice.
   (c) At a distance equal to about two orifice diameters downstream from the plane of the orifice.
   (d) At a distance equal to about one-half the orifice diameter downstream from the plane of the orifice.
   (e) At a distance within 10% of the orifice diameter downstream from the plane of the orifice.

The vena contracta is located about 1/2 orifice diameter downstream from the orifice. The reduced cross sectional area of the stream is expressed as a coefficient of contraction, $C_c = \dfrac{\text{area at vena contracta}}{\text{area of orifice}}$.

A coefficent of velocity, $C_v = \dfrac{\text{actual velocity at v.c.}}{\text{theoretical velocity at v.c.}}$.

Theoretical velocity at the vena contracta ignores the existence of an entrance loss to the orifice, and for a hole in a tank equals $\sqrt{2g_c h}$.
Coefficient of discharge, $C_D = C_c C_v$.

Answer is (d) $\bullet$

FLUIDS 54

An orifice 2" in diameter discharges fluid from a tank with a head of 15 feet. Discharge rate, q, is measured at 0.5 cfs. Actual velocity at the vena contracta, v.c., is 29.0 ft/sec. The coefficient of discharge, C_D, is nearest to:

(a) 0.62
(b) 0.74
(c) 0.79
(d) 0.86
(e) 0.94

$C_D = C_c C_v$ where C_D = coefficient of discharge

C_c = coeff. of contraction = $\dfrac{\text{area of v.c.}}{\text{area of orifice}}$

C_v = coeff. of velocity = $\dfrac{\text{actual velocity at v.c.}}{\text{theoretical velocity at v.c.}}$
(ignoring losses)

Theoretical velocity at v.c. = $V = \sqrt{2g_c h} = \sqrt{2(32.17)(15)}$ = 31.1 ft/sec

$C_v = \dfrac{29.0}{31.1} = 0.935$

Area of v.c. $Q = AV$ $A = \dfrac{Q}{V} = \dfrac{0.5}{29} = 0.0173 \ ft^2$

Area of orifice = $\dfrac{\pi}{4} D^2 = \dfrac{\pi}{4}\left(\dfrac{1}{6}\right)^2 = \dfrac{\pi}{144} = 0.0218 \ ft^2$

$C_c = \dfrac{0.0173}{0.0218} = 0.794$

$C_D = C_c C_v = (0.794)(0.935) = 0.742$ ●

Answer is (b)

FLUIDS 55

A stream of incompressible fluid leaves a nozzle at an absolute velocity of 50 ft/sec to the right. It enters, traverses and leaves tangent to a quarter circular blade which is moving at 30 ft/sec to the right as shown in the drawing. The magnitude of its absolute velocity in ft/sec after leaving the blade is nearest to:

(a) 24.
(b) 30.
(c) 36.
(d) 42.
(e) 48.

50 ft/sec

Ignoring frictional losses and any change of cross sectional area, fluid enters in line with blade travel at 20 ft/sec relative to the blade, and leaves normal to blade travel at 20 ft/sec relative to the blade.

Since the fluid also moves in the direction of blade motion at blade velocity, its absolute velocity as shown in the vector triangle is 36 ft/sec. ●

Fluid relative velocity + blade absolute velocity = fluid absolute velocity

Answer is (c)

FLUIDS 56

At normal atmospheric pressure, the maximum height in feet that a non volatile fluid of specific gravity of 0.80 may be siphoned is nearest to:

      (a) 12.
      (b) 20.
      (c) 28.
      (d) 34.
      (e) 42.

Maximum height to which a fluid may be siphoned is determined when pressure of the fluid column plus its vapor pressure equals external pressure. Minimum pressure at the highest point is 0 psia plus vapor pressure.

A 34 foot column of water has a pressure of 14.7 psi; this ignores its vapor pressure that is equivalent to 0.8 feet at 68 F.

$$(\text{ft of fluid})(62.4 \text{ lb/ft}^3 \times 0.80) = 14.7 \times 144 \text{ lb/ft}^2$$

$$\text{ft of fluid} = \frac{14.7 \times 144}{62.4 \times 80} = 42.4 \;●$$

Answer is nearest to (e)

FLUIDS 57

Cavitation is the result of
      (a) static pressure in a fluid becoming less than
            fluid vapor pressure
      (b) rivets under impact load
      (c) exposure of concrete to salt water
      (d) heat treatment of a low carbon steel
      (e) improper welding technique

Cavitation is characterized by local reduction of pressure within a fluid to the vapor pressure and the formation of a cavity within the flowing fluid. ●

Answer is (a)

FLUIDS 58

On doubling the speed, N, of a centrifugal pump, all of the following are true, except:

(a) Head, H is increased by a factor of 4.
(b) Horsepower, P is increased by a factor of 8.
(c) Volume flow rate, Q is increased by a factor of 2.
(d) Torque input is increased by a factor of 4.
(e) Head, horsepower and volume flow rate are independently variable.

Similarity relations are used between model and a geometrically similar prototype whose impeller diameters, D differ. These relations are:

Speed: $\dfrac{DN}{H^{1/2}}$ = constant $\dfrac{NQ^{1/2}}{H^{3/4}}$ = constant, specific speed

Discharge: $\dfrac{Q}{D^3N}$ = constant $\dfrac{Q}{D^2H^{1/2}}$ = constant

Power: $\dfrac{P}{D^5N^3}$ = constant

Applied to a single pump of fixed impeller diameter D, these reduce to

$$\frac{N_1}{N_2} = \frac{Q_1}{Q_2} = \frac{(H_1)^{1/2}}{(H_2)^{1/2}} = \frac{(P_1)^{1/3}}{(P_2)^{1/3}}$$

Doubling speed concurrently increases Q by a factor of 2,
increases H by a factor of 4, and
increases P by a factor of 8.

Mutual dependence of these variables complicates selection of pumps for specific applications.

The false answer is (e) ●

FLUIDS 59

Water flow rate in a 6 inch diameter pipe is measured with a differential pressure gage connected between a static pressure tap in the pipe wall and a pitot tube located at the pipe centerline. Which volume flow rate, q in cfs, results in a differential pressure of 1 psi?

(a)   0.2
(b)   2.4
(c)   15.6
(d)   28.2
(e) 344.0

A pitot tube generates a stagnation pressure as fluid kinetic energy is converted to pressure head.

Basis of calculation: 1 lb_m of fluid flowing

$$\frac{V^2}{2g_c} = \frac{P}{\rho} \qquad\qquad V = \left[\frac{2g_cP}{\rho}\right]^{1/2}$$

$$V = \left[\frac{2(32.17 \frac{lb_m \ ft}{lb_f \ sec^2})(\ 1 \ \frac{lb_f}{in^2})(\frac{144 \ in^2}{ft^2})}{62.4 \ \frac{lb_m}{ft^3}} \right]^{1/2} = 12.2 \ ft/sec$$

$$q = AV = \frac{\pi}{4} D^2 V = \frac{\pi}{4} \left(\frac{1}{2}\right)^2 (12.2) = 2.39 \ ft^3/sec \ \bullet$$

Answer is (b)

FLUIDS 60

The hydraulic formula $CA\sqrt{2gh}$ is used to find the

(a) quantity of discharge through an orifice
(b) velocity of flow in a closed conduit
(c) length of pipe in a closed network
(d) friction factor of a pipe
(e) wetted perimeter of a canal

For a static head orifice discharging freely into the atmosphere

$$Q = CA\sqrt{2gh} \ \bullet$$

Answer is (a)

FLUIDS 61

The hydraulic radius of an open-channel section is defined as:

(a) the wetted perimeter divided by the cross sectional area.
(b) the cross sectional area divided by the total perimeter.
(c) the cross sectional area divided by the wetted perimeter.
(d) one-fourth the radius of a circle with the same area.
(e) four times the radius of a circle with the same area.

Hydraulic radius R, in feet, is defined as $\dfrac{\text{cross sectional area, ft}^2}{\text{wetted perimeter, ft}}$ $\bullet$

For a full circular pipe: $R = \dfrac{\pi r^2}{2\pi r} = \dfrac{r}{2} = \dfrac{D}{4}$

Answer is (c)

FLUIDS 62

To calculate a Reynolds number for flow in open channels and in cross sections, one must utilize hydraulic radius, R in feet and modify the usual expression for circular cross sections:

$$Re = \frac{DV\rho}{\mu} = \frac{VD}{\upsilon}$$

where
D = diameter in feet
V = velocity in ft/sec
ρ = density in slugs/ft^3
μ = absolute viscosity in slugs/(ft sec)
υ = kinematic viscosity in ft^2/sec

Which of the following modified expressions for Re is applicable to flow in open or non-circular cross sections?

(a) $\dfrac{RD}{\upsilon}$

(b) $\dfrac{RV\rho}{\mu}$

(c) $\dfrac{2RD}{\upsilon}$

(d) $\dfrac{2RV\rho}{\mu}$

(e) $\dfrac{4RV}{\upsilon}$

Choices (a) and (c) are not dimensionless as required for Reynolds number.

Since hydraulic radius $R = \dfrac{\text{cross sectional area}}{\text{wetted perimeter}}$,

For a circular cross section:

$$R = \frac{\frac{\pi D^2}{4}}{\pi D} = \frac{D}{4}$$

Therefore,

$$Re = \frac{4RV\rho}{\mu} = \frac{4RV}{\upsilon} \quad \bullet$$

Answer is (e)

FLUIDS 63

In the design of waterways, the "hydraulic jump" is sometimes used for

(a) energy dissipation
(b) elimination of turbulence
(c) prevention of sedimentation
(d) measurement of flow
(e) reduction of head loss

Answer is (a) ●

FLUIDS 64

The rate of laminar water flow in a saturated soil can be calculated using:

(a) a Moody diagram.
(b) the Bernoulli equation.
(c) Darcy's law.
(d) the Hazen-Williams formula.
(e) the Chezy or Manning equations.

~~~~~~~~~~

A <u>Moody diagram</u> graphically presents the relationship for pipes among the dimensionless parameters: friction factor (f), Reynolds no. (Re), and surface roughness, $\frac{\epsilon}{D}$

The <u>Bernoulli equation</u> is a mechanical energy balance based on the 1st law of thermodynamics. It includes PE, KE, flow energy, losses and shaft work inputs, but omits thermal energy. The Euler equation is similar, though written in differential equation form.

The <u>Darcy-Weisbach formula</u> relates head loss (h_L) with friction factor (f), length-diameter ratio $\frac{L}{D}$ and kinetic energy $\frac{v^2}{2g}$.

$$h_L = f \left(\frac{L}{D}\right) \frac{v^2}{2g}$$

<u>Darcy's law</u> for diffusional flow closely parallels conductive heat transfer. As applied in this case:

$$Q = kiA \qquad \text{where } Q = \text{volume flow rate}$$
$$k = \text{coefficient of permeability}$$
$$i = \text{gradient}$$
$$A = \text{cross sectional area}$$

The <u>Hazen-Williams formula</u> is an empirical equation used in the solution of piping networks:

$$V = 1.318C_1 R^{0.63} S^{0.54}$$

where V = velocity in ft/sec
C_1 = a coefficient of relative roughness
R = hydraulic radius in ft
S = slope of the hydraulic gradient in ft lb_f/lb_m per ft length.

The <u>Chezy formula</u> applies to steady flow in open channels:

$$V = C \sqrt{RS}$$

where V = velocity
C = an empirical coefficient
R = hydraulic radius
S = slope of the hydraulic gradient (or slope of the free water surface in steady flow).

The <u>Manning equation</u> similarly applies to steady flow in open channels:

$$Q = AV = \frac{1.486}{n} AR^{2/3} S^{1/2}$$

> where Q = volume flow rate
> A = cross sectional area
> n = surface roughness factor
> R = hydraulic radius
> S = slope of the hydraulic gradient, as above.

Answer is (c) ●

FLUIDS 65

A rain storm occurred over a 100 square mile drainage area located in the Central Valley. The storm had a total duration of 3.0 hours and amounted to a total average rainfall over the drainage area of 1.00 inches of which 0.50 inch fell in the first hour, 0.30 inch in the second hour, and 0.20 inch in the third or last hour. Soil type A occurs in 30% of the area, Soil type B occurs in 25% of the area, and Soil type C occurs in 45% of the area.

The following data give the rainfall infiltration rates applicable to this 100 square mile drainage area:

	Rainfall Infiltration Rate*		
	Soil type A	Soil type B	Soil type C
First Hour	.40"	.35"	.30"
Second Hour	.25"	.20"	.15"
Third Hour	.20"	.15"	.10"

*Inches of rainfall absorbed by the soil per hour

Using the above data, compute the volume of storm runoff (cubic feet) that would originate from this 100 square mile drainage area.

〰〰〰〰〰

A tabular form simplifies the computation:

	Total Rain fall	Soil Type A 30 Sq Miles		Soil Type B 25 Sq Miles		Soil Type C 45 Sq Miles	
		Infl.	Run-off	Infl.	Run-off	Infl.	Run-off
First Hour	0.5	0.40	0.10	0.35	0.15	0.30	0.20
Second Hour	0.3	0.25	0.05	0.20	0.10	0.15	0.15
Third Hour	0.2	0.20	0	0.15	0.05	0.10	0.10
	1.0		0.15		0.30		0.45

Runoff = 30(0.15) + 25(0.30) + 45(0.45) = 32.25   sq mile - inches

Convert square miles to square feet and inches to feet

Runoff = 32.25 x $(5,280)^2$ x (1/12) = 75 x 10^6 cubic feet ●

FLUIDS 66

A concrete pipe is laid on a slope of 0.0001 foot per foot and is to carry 52.8 cubic feet per second of water when the pipe flows 0.75 diameter full.

With the use of Manning's formula and an "n" factor of 0.015, what size pipe should be used?

The Manning formula:

$$Q = \frac{1.486}{n} AR^{2/3} S^{1/2}$$

where n = roughness factor (0.015)
S = slope (0.0001)
Q = flow rate (52.8 cu ft/sec)
A = area of flow
R = hydraulic radius

Hydraulic radius $R = \frac{A}{P} = \frac{\text{Area of flow}}{\text{Wetted perimeter}}$

$$= \frac{\text{Area of circle - (Sector AOCE - Triangle AOC)}}{\text{Arc ABC}}$$

Angle $\Theta = \cos^{-1} (\frac{0.25D}{0.50D}) = \cos^{-1} 0.5 \qquad \Theta = 60°$

Area of Sector AOCE $= 2\left(\frac{60°}{360°}\right)\frac{1}{4}\pi D^2 = 0.262D^2$

Length of Arc ABC $= \pi D - 2\left(\frac{60°}{360°}\right)\pi D = 2.09D$

Area of Triangle AOC $= \frac{1}{2}(0.866)\left(\frac{D}{2}\right)(2)\left(\frac{D}{4}\right) = 0.108D^2$

$$R = \frac{A}{P} = \frac{0.785D^2 - (0.262D^2 - 0.108D^2)}{2.09D} = \frac{0.631D^2}{2.09D} = 0.302D$$

Substitute the values into the Manning equation

$$52.8 = \frac{1.486}{0.015}(0.631D^2)(0.302D)^{2/3}(0.0001)^{1/2}$$

$$D^{8/3} = 187.5$$

$$D = 7.12 \text{ feet} \ \bullet$$

7

THERMODYNAMICS

THERMO 1

Equations of state for a single component can be any of the following, <u>except</u>:

    (a)   the ideal gas law, $Pv = RT$.

    (b)   the ideal gas law modified by insertion of a compressibility factor, $Pv = ZRT$.

    (c)   any relationship interrelating 3 or more state functions.

    (d)   a mathematical expression defining a path between states.

    (e)   relationships mathematically interrelating thermodynamic properties of the material.

All <u>except</u> (d) are correct. The ideal gas law is the simplest equation of state; it is often applied to real gases by using a compressibility factor Z. Any relationships that interrelate thermodynamic state function data are equations of state. Item (d) expresses the path of a process between states rather than a relationship between variables at a single point or state.

Answer is (d) ●

THERMO 2

The state of a thermodynamic system is always defined by its

    (a)   absolute temperature

    (b)   process

    (c)   properties

    (d)   temperature and pressure

    (e)   availability

Availability is the amount of work that exists above complete equilibrium with ambient conditions. State is always defined by its properties. The number of properties required is determined by the Gibbs phase rule $P + F = C + 2$, where P = no. of phases, F = degrees of freedom or number of properties required, and C = no. of distinguishable components present.

Answer is (c) ●

THERMO 3

On the Mollier diagram for steam, which of the numbered lines represents a line of constant pressure?

    (a) Line 1
    (b) Line 2
    (c) Line 3
    (d) Line 4
    (e) Line 5

(a)   constant moisture 100 - x% or quality x%. (Line 1)

(b)   constant temperature, °F. (Line 2)

(c)   constant superheat, °F. (Line 3)

(d)   constant pressure, psia. (Line 4)

(e)   constant entropy, BTU/lb °R. (Line 5)

Horizontal lines are constant enthalpy or total heat content.

Answer is (d) ●

THERMO 4

On the Mollier diagram for steam (above) which lettered point represents the critical point at 3206 psia and 705°F?

The following data can be read for these points on a large Mollier diagram:

	Temp °F	P psia	Remarks	H BTU/lb	S BTU/lb °F
a	705	3206	critical point	910	1.065
b	162	5	22% moisture	910	1.490
c	141	3	10% moisture	1010	1.760
d	440	375	dry saturated	1205	1.490
e	780	200	superheated 400°	1413	1.760

Answer is (a) ●

THERMO 5

Mathematically, a thermodynamic property is which of the following?

    (a) a point function
    (b) a path function
    (c) an inexact differential
    (d) discontinuous
    (e) an exact differential

Thermodynamic properties, such as P, v, T, U, H, and S, are point functions, their derivatives are exact differentials. Heat, Q, and work, W, are path functions expressible as areas that are dependent upon path between points (states), their derivatives are inexact differentials.

Answer is (a) ●

THERMO 6

T	P psia	v_f	v_g	h_f	h_g	S_f	S_g
40°F	51.7	0.0116	0.774	17.3	81.4	0.0375	0.166
80°F	98.9	0.0123	0.411	26.4	85.3	0.0548	0.164
120°F	172.	0.0132	0.233	36.0	88.6	0.0717	0.162

Given the above data for Freon 12, what is its state at 40°F and 25 psia?

(a) saturated liquid
(b) superheated vapor
(c) compressed liquid
(d) saturated vapor
(e) vapor-liquid mixture

At 40°F equilibrium between liquid and gas exists at 51.7 psia. Below 51.7 psia superheated vapor exists, and above 51.7 psia only pressurized liquid exists.

Answer is (b) ●

THERMO 7

Using the previous Freon 12 data table, what is its entropy in BTU/lb°R at 120°F and 80% quality?

(a) 0.057
(b) 0.144
(c) 0.186
(d) 28.8
(e) none of these

At 120°F, S_f = 0.0717 and S_g = 0.162. S_{fg} = 0.162 - 0.0717 = 0.090

S_f is saturated liquid at 0% quality and S_g is saturated vapor of 100% quality.

S at 80% quality = S_f + (0.80)S_{fg} = 0.0717 + (0.80)(0.090) = 0.144 BTU/lb°R. ●

Answer is (b)

THERMO 8

Using the previous Freon 12 data table, what is its latent heat (heat of vaporization) in BTU/lb at 80°F?

(a) 0.219
(b) 0.423
(c) 26.4
(d) 58.9
(e) none of these

h_{fg} = h_g - h_f = 85.3 - 26.4 = 58.9 BTU/lb ●

Answer is (d)

THERMO 9

Normal boiling point of liquid oxygen is 90°K. What is this temperature in °R?

(a) - 330°R
(b) - 183
(c) 162
(d) 168
(e) 518

$$\left(\frac{1.8^\circ R}{^\circ K}\right)(90^\circ K) = 162^\circ R. \bullet$$

Answer is (c)

THERMO 10

For a typical refrigerant, draw a pressure-enthalpy diagram which shows the boundaries of solid, liquid, and vapor states. Identify on the diagram the following: (a) superheated vapor zone
(b) critical pressure
(c) compressed liquid zone
(d) saturated liquid line

THERMO 11

A cylinder fitted with a weightless, frictionless piston contains m pounds of air at temperature T_1, volume V_1, and ambient pressure P_a. Heat is added until the air in the cylinder has a temperature T_2, a volume V_2, and ambient pressure P_a. The specific heat of air at constant pressure is C_p, and the specific heat of air at constant volume is C_v. The heat transferred during the process is:

(a) $mC_p(T_2 - T_1)$

(b) $mC_v(T_2 - T_1) - P_a(V_2 - V_1)$

(c) $mC_p(T_2 - T_1) + P_a(V_2 - V_1)$

(d) $mC_v(T_2 - T_1) + P_a(V_2 - V_1)$

(e) $mC_v(T_2 - T_1)$

Consider one or more processes from $P_a V_1 T_1$ to $P_a V_2 T_2$ in a closed system and apply the first law of thermodynamics: $\Delta U = Q - W$.

For an ideal gas (air): $\Delta U = mC_v \Delta T$

Work done by a closed system at constant pressure: $W = \int_{V_1}^{V_2} PdV = P_a(V_2 - V_1)$

Combine these: $mC_v(T_2 - T_1) = Q - P_a(V_2 - V_1)$

Therefore $Q = mC_v(T_2 - T_1) + P_a(V_2 - V_1)$ ⬤ Answer is (d)

Dimensionality check:
Compatible units would be

$$BTU = (\text{1b mole})(\frac{BTU}{°F\text{-1b mole}})(°F) + (\frac{1bf}{ft^2})(ft^3)(\frac{BTU}{778\ 1bf\ ft})$$

THERMO 12

A nonflow (closed) system contains 1 lb of an ideal gas ($C_p = 0.24$, $C_v = 0.17$). The gas temperature is increased by $10°F$ while 5 BTU of work are done by the gas. What is heat transfer in BTU?

    (a)   -3.3
    (b)   -2.6
    (c)   +6.7
    (d)   +7.4
    (e)   none of these

The thermodynamic sign convention is + for heat in and + for work out of a system. Apply the first law for a closed system and an ideal gas working fluid:

$$\Delta U = mC_v \Delta T = Q - W$$

$$0.17(10) = Q - (+5) \qquad 1.7 = Q - 5, \qquad Q = 6.7 \ ⬤$$

Answer is (c)

THERMO 13

Shaft work of -15 BTU/lb and heat transfer of -10 BTU/lb change enthalpy of a system by

    (a)   -25 BTU/lb
    (b)   -15 BTU/lb
    (c)   -10 BTU/lb
    (d)   -5 BTU/lb
    (e)   +5 BTU/lb

The first law applied to a flow system is:

$$H = Q - W_s = -10 - (-15) = +5 \ ⬤$$

Answer is (e)

THERMO 14

The first law of thermodynamics states that:
    (a)   heat energy cannot be completely transformed into work.

    (b)   internal energy is due to molecular motion.

    (c)   heat can only be transferred from a body of higher termperature to one of lower temperature.

    (d)   energy can be neither created nor destroyed.

    (e)   entropy of the universe is increased by irreversible processes.

Item (a) is a consequence of the second law. Item (b) is a statement derived from the fact that internal energy is thermal kinetic energy of the molecules.

Item (c) is untrue as stated because a heat pump or refrigerator transfers heat from a low temperature reservoir to a higher temperature. Item (d) is correct, excluding nuclear processes that convert mass to energy. The first law is often cited as "conservation of energy." Item (e) is true, and is a consequence of the second law.

Answer is (d) ●

THERMO 15

A thermodynamic system originally contains a rock on a cliff and a pond of water below, both at the same temperature as the surroundings. The rock originally with PE = 4 BTU falls into the water and both ultimately come again into the same thermal equilibrium with the surroundings. Which row of data best represents the NET change in BTU that the rock plus pond system has undergone?

	ΔU	ΔKE	ΔPE	Q	W
(a)	0	0	-4	-4	0
(b)	+4	0	-4	0	0
(c)	0	+4	-4	0	0
(d)	0	0	0	-4	+4
(e)	-4	+4	-4	+4	0

Define the system as rock plus pond, with all else as surroundings. Initially rock has PE = +4 and KE = 0, and pond has PE = 0 and KE = 0, for a total = +4.

Finally rock has PE = 0 and KE = 0, and pond has PE = 0 and KE = 0, for a total of 0.

Energy was dissipated as heat, -Q to the surroundings. No <u>net</u> temperature change has occurred so ΔU = 0. No work was done by the system so W = 0.

Net result: ΔU = 0, ΔKE = 0, ΔPE = -4, Q = -4 and W = 0. ●

Answer is (a)

THERMO 16

A fluid at 100 psia has a specific volume of 4 ft^3/lb. and enters an apparatus with a velocity of 500 ft/sec. Heat radiation losses in the apparatus are equal to 10 BTU/lb. of fluid supplied. The fluid leaves the apparatus at 20 psia with a specific volume of 15 ft^3/lb. and a velocity of 1000 ft/sec. In the apparatus, the shaft work done by the fluid is equal to 195,000 ft-lb$_f$/lb$_m$.

Does the internal energy of the fluid increase or decrease, and how much is the change?

Basis of calculation: 1 lb$_m$

Q = -10 BTU

W_s = +195,000 ft lb$_f$

P_1 = 100 psia

v_1 = 4.0 ft^3/lb.

500 ft/sec

P_2 = 20 psia

v_2 = 15 ft^3/lb.

1000 ft/sec

Use the thermodynamic sign convention that heat in and work out are positive. First law energy balance for the flow system:

$$h_2 + KE_2 - h_1 - KE_1 = Q - W_s$$

Since the working fluid is unspecified and internal energy change is desired, use the definition: $h = u + Pv$.

$$u_2 + P_2v_2 + KE_2 - u_1 - P_1v_1 - KE_1 = Q - W_s$$

Or

$$u_2 - u_1 = Q - W_s + P_1v_1 + KE_1 - P_2v_2 - KE_2$$

Calculate numerical values for all terms except $u_2 - u_1$.

$$P_2v_2 = \frac{20(144)(15)}{778} = 55.5 \text{ BTU/lb} \qquad P_1v_1 = \frac{100(144)(4.0)}{778} = 74.0 \text{ BTU/lb}$$

$$KE_2 = \frac{v^2}{2gJ} = \frac{(1000)^2}{(64.4)(778)} = 20.0 \text{ BTU/lb}$$

$$KE_1 = \frac{v^2}{2gJ} = \frac{(500)^2}{(64.4)(778)} = 5.0 \text{ BTU/lb}$$

$$W_s = \frac{195,000 \text{ ft-lbf}}{lb_m} \quad \frac{BTU}{778 \text{ ft-lb}_f} = +250.6 \text{ BTU/lb}$$

Therefore,

$$u_2 - u_1 = -10 - (+250.6) + 74.0 + 5.0 - 55.5 - 20.0 = -257.1 \text{ BTU/lb} \quad \bullet$$
$$\text{decrease}$$

THERMO 17

An engine developing 30 brake horsepower must be cooled by a radiator through which water is pumped from the engine. Using the following factors:

Heat transferred from engine to cooling water, expressed as horsepower, equals 40 HP.

Water temperatures in radiator:
Top water = 200°F.
Bottom water = 190°F.

Weight of 1 gallon of water = 8.33 lbs.

Calculate the required rate of flow of water in gallons per minute.

〰〰〰〰

Basis of calculation: 1 minute.

Heat loss to cooling water = $40 \text{ HP} \left(\frac{2545 \text{ BTU}}{\text{HP-hour}}\right)\left(\frac{1 \text{ hr}}{60 \text{ min}}\right) = 1700 \text{ BTU/min}$

Heat loss = heat gain at steady state.

Heat gain by cooling water = 1700 BTU/min = $\dot{m}C_p\Delta T$, where $\dot{m}$ is water mass flow rate in lb/min.

$$\dot{m} = \frac{1700}{(1.0)(10)} = 170 \text{ lb. water/min.}$$

Volume flow rate = $\frac{170 \text{ lb.}}{\text{min.}} \times \frac{\text{gal.}}{8.33 \text{ lb.}} = 20.4 \text{ gal/min.} \quad \bullet$

THERMO 18

Saturated steam is supplied to a heat exchanger. 500 gallons per minute of water are being heated from 60° to 140°F in the exchanger. Condensate is being discharged to a receiver which is vented to the atmosphere.

A. If the entering steam is supplied to the heat exchanger at a pressure of 120 psia, and is condensing at this pressure:

      (1) How many pounds per hour of steam are used?

      (2) What percent of condensate is lost by flashing?

B. If the pressure of the steam supplied is reduced by a pressure reducing valve so that the steam is condensing at 30 psia, then:

      (3) How many pounds per hour of steam are used?

      (4) What percent of condensate is lost by flashing?

C. If fuel for the boiler supplying steam to the above heat exchanger costs 40¢ per million BTU's and overall boiler efficiency is 80 percent, which of the above systems would you use? Assume condensate is returned to the boiler at 212°F and make-up water is at 60°F.

ENTHALPY

	14.7 psia	30 psia	120 psia
Saturated Liquid, h_f	180.1	218.8	312.5
Evaporation, h_{fg}	970.3	945.2	878.1
Saturated Vapor, h_g	1150.4	1164.0	1190.6

Cooling Water ⟶ | ⟶ Cooling Water at 500 gpm ($\Delta T = 80°F$)

Dry Saturated Steam 120 psia

Saturated Liquid 120 psia

Condensate at 1 atm.

Saturated Steam at 14.7 psia

Saturated Liquid at 14.7 psia

A. (1) <u>Calculate steam flow rate</u> (Basis: 1 hour)

Water mass flow rate, $\dot{m}_1 = \dfrac{500 \text{ gal}}{\text{min}} \times \dfrac{60 \text{ min}}{\text{hour}} \times \dfrac{8.33 \text{ lb.}}{\text{gal.}} = 250,000 \dfrac{\text{lb.}}{\text{hour}}$

Heat balance: heat to water = heat from condensing steam

$\dot{m}_1 C_p (T_2 - T_1) = \dot{m}_2 h_{fg}$, where $\dot{m}_2$ is steam mass flow rate, lb/hr.

From saturated steam table at 120 psia, h_{fg} = 878.1 BTU/lb.

$\dot{m}_2 = \dfrac{\dot{m}_1 C_p (T_2 - T_1)}{h_{fg}} = \dfrac{250,000(1)(140 - 60)}{878.1} = 22,800 \dfrac{\text{lb.}}{\text{hour}}$ ●

A. (2) <u>Percent condensate lost by 120 psia flashing</u> (Basis: 1 lb.)

Mass balance: X = Weight fraction flashed off.

1 lb. sat. liquid at 120 psia = (X)lb. sat. vapor at 14.7 psia
                                + (1 - X) lb. sat. liquid at 14.7 psia

Heat balance:

$1(h_f$ at 120 psia) $= X(h_g$ at 14.7 psia) $+ (1 - X)(h_f$ at 14.7 psia)

$1(312.5) = X(1150.4) + (1 - X)(180.1)$

$312.5 = 1150.4X + 180.1 - 180.1X$

$X = 0.1365$ Therefore 13.65% of condensate lost by flashing ●

B. Throttle steam to 30 psia (a constant enthalpy process)

During throttling: $h_{in} = h_{out}$

Saturated steam at 120 psia ($h_g = 1190.6$) becomes superheated steam at
                30 psia ($h_g = 1190.6$).

From saturated steam table at 30 psia, $h_f = 218.8$ BTU/lb.

Heat liberated from steam: $1190.6 - 218.8 = 971.8$ BTU/lb.

(3) <u>Calculate steam flow rate</u> (Basis: 1 hour)

Use heat balance as in (1) above: heat to water = heat from steam

$$\dot{m}_1 C_p (T_2 - T_1) = \dot{m}_2 (971.8)$$

$$\dot{m}_2 = \frac{250,000(1)(140 - 60)}{971.8} = 20,600 \text{ lb./hour} ●$$

(4) <u>Percent condensate lost by 30 psia flashing</u> (Basis: 1 lb.)

Mass balance: Y = Weight fraction flashed off

1 lb. sat. liquid at 30 psia = (Y)lb. sat. vapor at 14.7 psia
                               + (1 - Y) lb. sat. liquid at 14.7 psia

Heat balance:

$1(h_f$ at 30 psia) $= Y(h_g$ at 14.7 psia) $+ (1 - Y)(h_f$ at 14.7 psia)

$1(218.8) = Y(1150.4) + (1 - Y)(180.1)$

$Y = 0.0399$ Therefore 3.99% of condensate lost by flashing ●

C. <u>Proposal Comparison</u>

	120 psia condensation and flashing	30 psia condensation and flashing
Steam rate, lb./hour	22,800.	20,600.
Steam and Condensate lost, lb/hour	3,120.	820.
Condensate returned, lb/hour	19,680.	19,780.

Superficially, it is apparent that the 30 psia case loses much less heat to
the environment by flashing. Limitation on system operability is imposed
by temperature on the steam side of the heat exchanger. Temperature of
120 psia saturated steam is 341°F, and superheated steam at 30 psia
(h = 1190.6) is determined from superheated steam tables to be 303°F. The
lower temperature requires more condenser area.

The magnitude of the heat saving is shown by an overall heat balance.

Overall heat balance on water-steam side of the boiler (Basis: 1 hour)

```
┌─────────────────┐
│ Water-Steam     │──────────▶  22,800 lbs. saturated steam at 120 psia (1190.6 BTU/lb.)
│ side of         │
│ boiler          │
│ (120 psia case) │◀──────────  19,680 lbs. condensate at 212°F (180 BTU/lb.)
└─────────────────┘
      ▲       ▲
      │       │     3120 lbs. makeup at 60°F
      Q       │     (28 BTU/lb, above 32°F reference of steam tables).
   BTU/hr.
```

heat in = heat out

$$Q + 3120(28) + 19,680(180) = 22,800(1190.6)$$

$$Q + 0.09 \times 10^6 + 3.54 \times 10^6 = 27.15 \times 10^6$$

$$Q = 23.52 \times 10^6 \text{ BTU/hour}$$

```
┌─────────────────┐
│ Water-Steam     │──────────▶  20,600 lbs. saturated steam at 120 psia (1190.6 BTU/lb.)
│ side of         │
│ boiler          │
│ (30 psia case)  │◀──────────  19,780 lbs. condensate at 212°F (180 BTU/lb.)
└─────────────────┘
      ▲       ▲
      │       │     820 lbs. makeup at 60°F
      Q       │     (28 BTU/lb.)
   BTU/hr.
```

heat in = heat out

$$Q + 820(28) + 19,780(180) = 20,600(1190.6)$$

$$Q + 0.02 \times 10^6 + 3.56 \times 10^6 = 24.55 \times 10^6$$

$$Q = 20.97 \times 10^6 \text{ BTU/hour}$$

Summary

Heat delivered to the water-steam side costs $\frac{\$0.40}{.80}$ = \$0.50/10^6 BTU and the 30 psia case offers a fuel economy, but the condenser heat transfer area requirement may preclude its feasibility. A better thermal approach would be to return 120 psia sat'd liq. from condenser to boiler. ●

THERMO 19

55,000 gallons of water passes through a heat exchanger and absorbs 28,000,000 BTU's. The exit temperature is 110°F. The entrance water temperature in °F is nearest to:

(a) 49.
(b) 56.
(c) 68.
(d) 73.
(e) 82.

〰〰〰〰〰

C_p for liquid water = 1.0 BTU/(lb.°F), density = 8.33 lb./gallon.

$$Q = mC_p \Delta T = mC_p(T_2 - T_1)$$

$$28,000,000 = (55,000 \text{ gal.})(\frac{8.33 \text{ lb.}}{\text{gallon}})(1.0)(110 - T_1)$$

$$61.2 = 110 - T_1 \qquad T_1 = 48.8°F ●$$

Answer is (a)

THERMO 20

The mass flow rate of Freon 12 through a heat exchanger is 10 pounds/minute. Enthalpy of Freon entry is 102 BTU/lb. and of Freon exit is 26 BTU/lb. Water coolant is allowed to rise $10°F$. The water flow rate in pounds/minute is:

(a) 24.
(b) 76.
(c) 83.
(d) 112.
(e) 249.

Basis of calculation: 1 minute

Heat gain by water - Heat lost by Freon = 0

$$m_1 C_p \Delta T - m_2(h_2 - h_1) = 0$$

$$m_1(1)(10) - 10(26 - 102) = 0 \qquad m_1 = \frac{760}{10} = 76. \text{ lbs./min.} \; \bullet$$

Answer is (b)

THERMO 21

Five gallons per minute of hot water at $180°F$ is produced in a flow system by injection and condensation of low pressure steam at 20 psi (gage) and 80% quality into cold water at $60°F$. Calculate the steam flow rate in pounds per minute.

Calculation basis: 1 minute

(1) First law energy balance, ignoring any PE and KE changes, for the flow system.

$$\dot{m}_3 h_3 - \dot{m}_1 h_1 - \dot{m}_2 h_2 = Q - W_s = 0 \quad \text{since no heat is lost or work done.}$$

(2) Mass flow rate balance: $\dot{m}_1 = \dot{m}_3 - \dot{m}_2$

$$\dot{m}_3 = \frac{5 \text{ gal}}{\text{min}} \; \frac{8.33 \text{ lb.}}{\text{gal.}} = 41.67 \text{ lb./min}$$

A more precise calculation would utilize saturated steam table data at $180°F$ where $v_f = 0.01605 \text{ ft}^3/\text{lb}$.

$$\dot{m}_3 = \frac{5 \text{ gal}}{\text{min}} \; \frac{\text{ft}^3}{7.48 \text{ gal}} \times \frac{\text{lb.}}{0.01605 \text{ ft}^3} = 41.7 \text{ lb./min}$$

(This allows for the trivial thermal expansion of water.)

(3) Determine enthalpies from saturated steam tables:

h_1 = enthalpy of liq. water at $60°F$ = h_f = 28.07 BTU/lb.

h_3 = enthalpy of liq. water at $180°F$ = h_f = 147.91 BTU/lb.

h_2 = enthalpy of 34.7 psia (20 + 14.7) steam of 80% quality:

Saturated Steam Table Data:

psia	h_f	h_{fg}	h_g
30	218.83	945.2	1164.0
40	236.02	933.7	1169.7

Double interpolation is required: for 80% quality, and for 34.7 psia pressure.

at 80% quality, (30 psia) h = 218.83 + 0.80(945.2) = 975.0 BTU/lb.
(40 psia) h = 236.02 + 0.80(933.7) = 983.0 BTU/lb.

Linearly interpolate between 30 and 40 for 34.7 psia at 80% quality:

$$h_2 = 975.0 + \frac{4.7}{10}(983.0 - 975.0) = 978.8 \text{ BTU/lb.}$$

(4) Solve energy balance for $\dot{m}_2$, using $\dot{m}_1 = \dot{m}_3 - \dot{m}_2$

$$\dot{m}_3 h_3 = \dot{m}_1 h_1 + \dot{m}_2 h_2$$

$$(41.7)(147.91) = (41.7 - \dot{m}_2)(28.07) + \dot{m}_2(978.8)$$

$$6167.8 = 1170.5 - 28.07\dot{m}_2 + 978.8\dot{m}_2$$

$$\dot{m}_2 = \frac{4997}{950.7} = 5.26 \text{ lb/min.} \quad \bullet$$

THERMO 22

Exhaust steam from a turbine exhausts into a surface condenser at a mass flow rate of 8000 lb/hr, 2 psia and 92% quality. Cooling water enters the condenser at 74°F and leaves at the steam inlet temperature. What is the cooling water mass flow rate in lb/hr?

〰〰〰〰〰

Saturated steam table data at 2 psia are:

T °F	h_f, BTU/lb	h_{fg}, BTU/lb	h_g, BTU/lb
126.08	93.99	1022.2	1116.2

h_1 = enthalpy of steam at 92% quality = $h_f + 0.92 h_{fg}$

= 94.0 + 0.92(1022.2) = 1034.4 BTU/lb.

h_2 = enthalpy of liquid water at 126.1°F = 94.0 BTU/lb.

h_3 = enthalpy of liquid water at 74°F = 42.0 BTU/lb above reference of 32°F

Alternate: find h_f in steam tables at 74°F ignoring negligible effect of pressure on enthalpy of incompressible liquids.

STEAM
2 psia, x = 0.92
$\dot{m}_1$ = 8000 lb/hr

126.1°F

74°F COOLING WATER
$\dot{m}_2$ = mass flow rate

In the absence of data, assume steam condensate leaves at 126.1°F

Heat balance, basis: 1 hour

heat from steam = heat to cooling water

$$\dot{m}_1(h_1 - h_2) = \dot{m}_2(h_2 - h_3)$$

$$8000(1034.4 - 94.0) = \dot{m}_2(94.0 - 42.0) \qquad \dot{m}_2 = 144,700. \text{ lb/hr} \;\bullet$$

THERMO 23

53.8 grams of finely divided aluminum is heated to 98.3°C and dropped into 76.2 grams of water at 18.6°C contained in a calorimeter. The final temperature of the mixture is 27.4°C. The mass of the calorimeter is 123 grams, and its specific heat may be taken as 0.092 cal $g^{-1}deg^{-1}$. The combined thermal capacity of the thermometer and metal stirrer is 6.5 cal deg^{-1}.

Assuming no heat is lost from the system, calculate the mean specific heat of aluminum for the above temperature range.

$$m_{Al}C_{Al}(T_1 - T_2) = \left[m_{cal}C_{cal} + m_{thermo}C_{thermo} + m_{H_2O}C_{H_2O} \right] (T_2 - T_0)$$

$$53.8C_{Al}(98.3 - 27.4) = \left[123(0.092) + 6.5 + 76.2(1) \right] (27.4 - 18.6)$$

$$3815C_{Al} = (11.31 + 6.5 + 76.2)(8.8)$$

$$C_{Al} = \frac{(94.01)(8.8)}{3815} = 0.217 \text{ cal/(gm °C)} \quad \bullet$$

THERMO 24

3000 cfm of 65°F air is required to maintain a house at 76°F.

25%wt of air entering the air conditioner is from outside at 90°F, and the remainder is recycled from inside at 76°F.

REQUIRED: Determine the rating in tons of the air conditioner.
(Assume sensible heat changes only, ignore any dehumidification.)

65°F Air Density: $\quad \rho_{65} = \dfrac{P}{RT} = \dfrac{14.7 \times 144}{53.3 \times 525 \text{ R}} = 0.0756 \text{ lbm/ft}^3$

Total mass flow rate $\dot{m}_T = \rho Q = 0.0756(3000) = \quad 227 \text{ lbm/min}$
$$= 13,620 \text{ lbm/hr}$$

$\dot{m}_{90}$ = mass flow rate 90° Air = 0.25(13,620) = 3405 lbm/hr

$\dot{m}_{76}$ = mass flow rate 76° Air = 0.75(13,620) = 10,215 lbm/hr

C_p = specific heat of Air = 0.24 BTU/lbm°F

$Q_c = \dot{m}_{90}C_p \Delta T + \dot{m}_{76}C_p \Delta T = 3405(0.24)(90° - 65°) + 10,215(0.24)(76° - 65°)$
$$= 20,430 + 26,970 = 47,400 \text{ BTU/hr}$$

1 ton of refrigeration = 200 BTU/min = 12,000 BTU/hr

$$\text{Tons of Air Conditioning} = \frac{47,400}{12,000} = 3.95 \quad \bullet$$

THERMO 25

In terms of Q_H (heat from high temperature source) and Q_L (heat to low temperature sink), the net work of a Carnot cycle is:

(a) $W = Q_H - Q_L$

(b) $W = \dfrac{Q_H - Q_L}{Q_L}$

(c) $W = \dfrac{Q_L}{Q_H - Q_L}$

(d) $W = Q_L - Q_H$

(e) none of these

The work produced by a carnot cycle heat engine is $Q_H - Q_L$ ●

Answer is (a)

THERMO 26

Select the relationship defining absolute temperature in terms of Q_H and Q_L, heats transferred in a Carnot cycle, and T_H and T_L, the absolute temperatures of the reservoirs involved.

(a) $\dfrac{Q_H - Q_L}{T_H - T_L} = 1$

(b) $\dfrac{Q_H - Q_L}{Q_H} = \dfrac{T_H}{T_L}$

(c) $\dfrac{Q_H}{Q_L} = \dfrac{T_H}{T_L}$

(d) $\dfrac{Q_H}{Q_L} = \dfrac{T_L}{T_H}$

(e) $\dfrac{Q_H - Q_L}{Q_H} = \dfrac{T_H - T_L}{T_H}$

Statement (c) defines absolute thermodynamic temperature. Statement

(e) is numerically equal to efficiency η of a Carnot cycle heat engine.

(c) is correct, and (e) is derived therefrom. ●

THERMO 27

The maximum thermal efficiency that can be obtained in an ideal reversible heat engine operating between $1540°F$ and $340°F$ is closest to

(a) 100%
(b) 60%
(c) 78%
(d) 40%
(e) 22%

$$\eta_{thermal} = \frac{W}{Q_H} = \frac{Q_H - Q_L}{Q_H}$$

$T_L = 340°F + 460° = 800°R$

$T_H = 1540°F + 460° = 2000°R$

$$= 1 - \frac{Q_L}{Q_H} = 1 - \frac{T_L}{T_H}$$

$$= 1 - \frac{800}{2000} = 1 - 0.40 = 0.60 = 60\% ●$$

Answer is (b)

THERMO 28

A 3 HP refrigerator or heat pump operates between 0°F and 100°F.
The maximum theoretical heat that can be transferred from the cold reservoir is
nearest to:

    (a)   7,600 BTU/hr
    (b) 13,000 BTU/hr
    (c) 23,000 BTU/hr
    (d) 35,000 BTU/hr
    (e) 43,000 BTU/hr

Coefficient of Performance of a refrigerator or heat pump:

$$C.O.P. = \frac{T_L}{T_H - T_L} = \frac{460°R}{560° - 460°} = 4.6$$

$$C.O.P. = \frac{Q_L}{W} = 4.6 \qquad 1 \text{ HP} = 2545 \text{ BTU/hr}$$

$$\therefore W = 7635 \text{ BTU/hr}$$

$$C.O.P. = \frac{Q_L}{Q_H - Q_L} = \frac{Q_L}{W} = \frac{Q_L}{7635}$$

$$Q_L = 35,100 \text{ BTU/hr} \ \bullet$$

Answer is (d)

THERMO 29

A Carnot cycle heat engine operating between 1540°F and 440°F has an
efficiency of approximately

    (a) 55%
    (b) 45%
    (c) 35%
    (d) 29%
    (e) 82%

For a Carnot cycle heat engine, efficiency $\eta = \dfrac{T_H - T_L}{T_H - 0} = \dfrac{2000°R - 900°R}{2000°R}$

$$= 0.55 \ \bullet$$

Answer is (a)

THERMO 30

Second law limitation on the maximum horsepower output from any power unit
burning 1,000,000 BTU/hr of fuel with high and low temperature extremes of
1540°F and 40°F is:

    (a) 98.
    (b) 295.
    (c) 1140.
    (d) 3830.
    (e) none of these

Basis of calculation: 1 hour

Carnot efficiency $\eta = \dfrac{T_H - T_L}{T_H - 0} = \dfrac{2000°R - 500°R}{2000°R} = 0.75 = \dfrac{Q_H - Q_L}{Q_H - 0} = \dfrac{W}{Q_H}$

$$W = 0.75Q_H = 0.75 \times 10^6 = 750,000 \text{ BTU/hr}$$

$$= \frac{750,000 \text{ BTU/hr}}{2545 \text{ BTU/(HP hr)}} = 295. \text{ HP} \bullet$$

Answer is (b)

THERMO 31

A heat pump installation is used to warm a stream of circulating air in a residence to 110°F. An outside ambient of 35°F is considered the available heat source. A heating requirement of 90,000 BTU/hr is necessary to maintain a comfortable temperature in the living space.

(a) Determine the absolute minimum electrical requirement in kilowatts and in HP to operate a heat pump delivering 90,000 BTU/hr under these conditions.

(b) Determine the electrical requirement in Kw to accomplish the same heat by resistance heating.

〰〰〰〰〰〰〰

A Carnot reversed cycle would provide the most efficient heat pump possible, although this is only approached in practice with Rankine vapor compression cycles.

T_H ── Area = W, work added

T_L ── Area = Q_L, heat from low temperature reservoir

T = 0 ── Both areas = Q_H, heat rejected at high temperature

For a heat pump: $Q_H = Q_L + W$ as shown in the T-S diagram

Coefficient of Performance:

$$C.O.P. = \frac{Q_L}{W} = \frac{T_L}{T_H - T_L}$$

Absolute temperatures are required.

$$110°F = 570°R, \quad 35°F = 495°R.$$

(a)

$$C.O.P. = \frac{Q_L}{W} = \frac{495}{570 - 495} = 6.60$$

$$Q_H = 90,000 = Q_L + W = 7.60 \ W$$

$$W = 11,840 \text{ BTU/hr} = \frac{11,840 \text{ BTU/hr}}{3413 \text{ BTU/Kwh}} = 3.47 \text{ Kw} \bullet$$

$$\frac{3.47 \text{ Kw}}{0.746 \text{ Kw/HP}} = 4.65 \text{ HP} \bullet$$

(b)

Resistance heating is pure energy dissipation

$$\frac{90,000}{3,413} = 26.4 \text{ Kw} \bullet$$

THERMO 32

In any non-quasistatic thermodynamic process, the overall entropy of
an isolated system will:

    (a) Increase and then decrease
    (b) Decrease and then increase
    (c) Stay the same
    (d) Increase only
    (e) Decrease only

quasistatic: infinitely slow, lossless, hypothetical, by differential
increments

The overall entropy will increase for an isolated system, or for the
system plus surroundings. ●

Answer is (d)

THERMO 33

Entropy is the measure of

    (a) the change in enthalpy of a system
    (b) the internal energy of a gas
    (c) the heat capacity of a substance
    (d) randomness or disorder
    (e) the total heat content of a system

Answer is (d) ●

THERMO 34

A quantity of working fluid undergoes three sequential irreversible processes:
    (1) Receives heat from high temperature reservoir (at high ΔT)
        Reservoir entropy decreases 0.240 BTU/°R and
        Fluid entropy increases 0.245 BTU/°R.

    (2) Does work in an adiabatic turbine.
        Fluid entropy increases 0.013 BTU/°R.

    (3) Discharges heat to low temperature reservoir (at high ΔT)
        Fluid entropy decreases 0.258 BTU/°R.
        Reservoir entropy increases 0.263 BTU/°R.

Entropy change for the universe is BTU/°R is:

    (a) indeterminate
    (b) -0.023
    (c) +0.010
    (d) +0.023
    (e) none of these

Net increase for the universe equals + .023, the difference between high
temperature reservoir loss and low temperature gain. This is a consequence
of the irreversibilities because the heat transfer is not at $\Delta T = 0$, and
there is lost work in the turbine. ●

Answer is (d)

THERMO 35

If heat transfers of the previous problem were reversible, and if the turbine
operated isentropically, what is the entropy gained or lost by the low
temperature reservoir?

    (a) indeterminate
    (b) 0
    (c) +0.240 BTU/°R
    (d) +0.250 BTU/°R
    (e) none of these

Reservoirs are considered infinite and no temperature rise occurs. Since ΔS for reversible heat transfer = 0, and the turbine is isentropic (ΔS = 0), entropy loss by the high temperature reservoir matches the entropy gain by the low temperature reservoir = +0.240 BTU/°R. ●

Answer is (c)

THERMO 36

A Carnot cycle heat engine operates between 1540°F and 40°F and rejects 250. BTU/lb to the low temperature reservoir or heat sink. Maximum entropy change in BTU/lb°R of the working fluid for any process in the cycle is:

(a) 0
(b) 0.25
(c) 0.50
(d) 1.0
(e) 4.0

The T-S diagram for a Carnot heat engine shows its relation between heat temperature and entropy.

Both areas = heat intake from high temperature reservoir = Q_H.

Since $250 = 500 \cdot \Delta S$ $\therefore \Delta S = 0.50$ ●

Answer is (c)

THERMO 37

For spontaneously occurring natural processes in an isolated system, which expression best expresses dS?

(a) $dS = \dfrac{dQ}{T}$

(b) $dS = 0$

(c) $dS > 0$

(d) $dS < 0$

(e) $dS = C_p\dfrac{dt}{T} - R\dfrac{dp}{P}$

(a) $dS = \dfrac{dQ_{rev}}{T}$ only. The reversible requirement is necessary to generate the exact height vs. rectangular area equivalence on the Carnot cycle T-S diagram.

(b) Only a reversible adiabatic process is isentropic by definition.

(c) All naturally occurring spontaneous processes are irreversible and result in an entropy increase. ●

(d) An energy input from surroundings is required to reduce entropy.

(e) This is an expression for entropy change in an ideal gas.

Answer is (c)

THERMO 38

Which of the following statements about entropy is <u>false</u>?

   (a) Entropy of a mixture is greater than that of its components under the same conditions.
   (b) An irreversible process increases entropy of the universe.
   (c) Entropy has the units of heat capacity.
   (d) Net entropy change in any closed cycle is zero.
   (e) Entropy of a crystal at $0°F$ is zero.

All are true except (e). The entropy of a perfect crystal at absolute zero ($0°K$ or $0°R$) is zero. This is the 3rd law of thermodynamics. There is presumably no randomness existent at this temperature in a crystal without flaws, impurities or dislocations.

Answer is (e) ●

THERMO 39

Work or energy can be a function of all of the following <u>except</u>:

   (a) force and distance
   (b) torque and angular rotation
   (c) power and time
   (d) force and time
   (e) temperature and entropy

   (a) $dW = FdS$ for translation

   (b) $dW = Td\theta$ for rotation

   (c) $\dfrac{dW}{dt} = power$ $\quad \therefore dW = (Power)dt$

   (d) $Fdt = d(mV)$, the impulse momentum equation. Impulse, Fdt, is not an energy term.

   (e) $dQ_{rev} = TdS$. Heat is a form of energy

All <u>except</u> (d) are correct ●

THERMO 40

Energy changes are represented by all <u>except</u> which one of the following?

   (a) $-\int VdP$
   (b) $TdS - PdV$
   (c) $TdS + VdP$
   (d) $\dfrac{dQ_{rev}}{T}$
   (e) $mC_p dT$

   (a) is shaft work.

   (b) $dU = TdS - PdV = dQ_{rev} - W$. dU is an internal energy change.

   (c) $dH = TdS + VdP$. dH is an enthalpy change.

   (d) $\dfrac{dQ_{rev}}{T} = dS$. dS is change in the thermodynamic state function, entropy. It is not an energy change.

   (e) $dH = mC_p dT$ for an ideal gas.

Therefore (d) represents an entropy or state change, all others represent energy changes. ●

Answer is (d)

THERMO 41

A high velocity flow of gas at 800 ft/sec possesses kinetic energy nearest to which of the following?

  (a)   1.03 BTU/lb
  (b)   4.10 BTU/lb
  (c)   9.95 BTU/lb
  (d) 12.8 BTU/lb
  (e) 41.0 BTU/lb

Basis: 1 lb_m of fluid flowing

$KE = \dfrac{V^2}{2g_c}$ in ft-lb_f, where V is in ft/sec, and $g_c = 32.17$

Use J = 778 ft-lb_f/BTU to convert to BTU.

$KE = \dfrac{800^2}{2(32.17)(778)} = 12.8$ BTU/lb. ●

Answer is (d)

THERMO 42

(U + PV) is a quantity called

  (a) flow energy
  (b) shaft work
  (c) entropy
  (d) enthalpy
  (e) internal energy

Flow energy is PV. Shaft work, W_s, is $-\int VdP$. Entropy is S.

Internal energy is U. Enthalpy H is defined as U + PV, the sum of internal energy plus flow energy. ●

Answer is (d)

THERMO 43

In flow process, neglecting KE and PE changes, $-\int VdP$ represents which item below?

  (a) heat transfer
  (b) shaft work
  (c) closed system work
  (d) flow energy
  (e) enthalpy change

Shaft work is work or mechanical energy crossing the fixed size boundary (control volume) of a flow (open) system. Shaft work W_s is defined, in the absence of PE and KE changes, by dH = TdS + VdP, where TdS = dQ_{rev} and -VdP is dW_s. In integrated form $\Delta H = \int TdS + \int VdP = Q_{rev} - W_s$, where W_s is represented by $-\int VdP$. Closed system work W is defined by dU = TdS - PdV, or $\Delta U = \int TdS - \int PdV = Q_{rev} - W$. Thus closed system work is $+\int PdV$.

Flow energy is the PV term, and enthalpy change is ΔH.

Answer is (b) ●

THERMO 44

Power may be expressed in units of

        (a) ft-lbs
        (b) BTU/hr
        (c) HP-hours
        (d) Kw-hours
        (e) BTU

Ft-lb, HP-hours, BTU, and Kwh are usual mechanical, thermal and electrical energy terms. Power = energy/time. Usual power units are ft-lb/sec, HP, BTU/hr and Kw.

Answer is (b) ●

THERMO 45

Given the following data: Electricity cost $0.015/Kwh
                                 Natural gas cost $0.065/100 cu ft
                                 Heat content of gas 1050 BTU/cu ft

Required: Compute how many times as expensive it would be to heat a house by electricity than to heat the same house by gas, if the electric heat is assumed to be 100% efficient and the gas is 60% efficient.

We will need to convert foot-pounds to BTU or vice versa

              1 BTU = 778 foot-pounds

Heating Cost:

$$\text{Natural Gas} = \frac{\$0.065}{0.60 \times 1050 \times 100} = \$1.03 \times 10^{-6}/BTU$$

$$\text{Electricity} = \frac{\$0.015 \times 0.746 Kw/HP \times 1\ HP/33,000\ ft\text{-}lb/min \times 1/60 \times 778}{1.0}$$

$$= \$4.40 \times 10^{-6}/BTU$$

$$\text{Ratio of } \frac{\text{Cost of electrical heating}}{\text{cost of gas heating}} = \frac{\$4.40 \times 10^{-6}/BTU}{\$1.03 \times 10^{-6}/BTU} = 4.27 \quad ●$$

THERMO 46

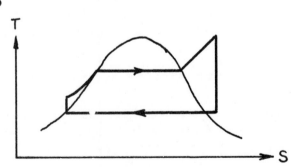

The above temperature-entropy diagram represents:

        (a) Rankine cycle with superheated vapor
        (b) Carnot cycle
        (c) Diesel cycle
        (d) Refrigeration cycle
        (e) Adiabatic process

Answer is (a) ●

THERMO 47

A Carnot heat engine cycle is represented on the T-S and P-V diagrams below:

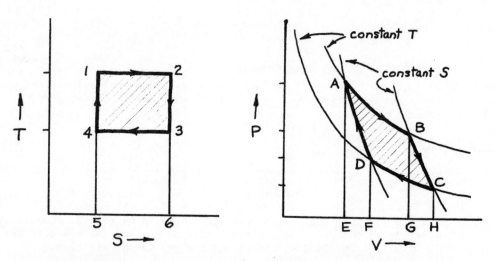

Which of the several areas bounded by numbers or letters represents the amount of heat rejected by the fluid during one cycle?

    (a)   area 1-2-6-5
    (b)   area B-C-H-G
    (c)   area 3-4-5-6
    (d)   area D-A-E-F
    (e)   area C-D-F-H

The table below gives the significance of each area of the diagrams:

PROCESS	T-S Diagram Area Representing Heat	P-V Diagram Area Representing Work
isothermal expansion 1-2 and A-B	1-2-6-5 = heat in from high temp. reservoir	A-B-G-E = work done by fluid
isentropic expansion 2-3 and B-C	2-3-6 = 0 heat transfer	B-C-H-G = work done by fluid
isothermal compression 3-4 and C-D	3-4-5-6 = heat out to low temp. reservoir	C-D-F-H = work done on fluid
isentropic compression 4-1 and D-A	4-1-5 = 0 heat transfer	D-A-E-F = work done on fluid
net result of process	1-2-3-4 = net heat converted to work	A-B-C-D = net work done by process

Answer is (c) ●

THERMO 48

Which of the following thermodynamic cycles is the most efficient?

    (a)   Carnot
    (b)   Brayton
    (c)   Otto
    (d)   Diesel
    (e)   Rankine

Answer is (a) ●

THERMO 49

A Carnot engine operating between $70°F$ and $2000°F$ is modified solely by raising the high temperature by $150°F$ and raising the low temperature by $100°F$. Which of the following statements is <u>false</u>?

    (a) thermodynamic efficiency is increased.
    (b) more work is done during the isothermal expansion.
    (c) more work is done during the isentropic compression.
    (d) more work is done during the reversible adiabatic expansion.
    (e) more work is done during the isothermal compression.

Carnot cycle efficiency is originally

$$\eta = \frac{T_H - T_L}{T_H - 0} = \frac{2460°R - 530°R}{2460°R} = 0.785$$

After the change

$$\eta = \frac{2610 - 630}{2610} = 0.76 \quad \text{Efficiency is therefore reduced.}$$

On the T-S and P-V diagrams below the original cycle is shown as ABCD, the modified cycle as A'B'C'D'.

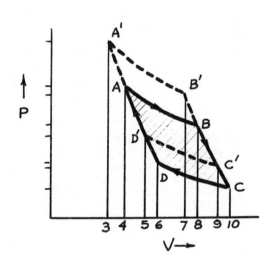

Compare work done during isothermal expansion (A to B, vs. A' to B')
    Original: area A-B-8-4
    Modified: area A'-B'-7-3 is larger

Compare work done during isentropic compression (D to A, vs. D' to A')
    Original: area D-A-4-6
    Modified: area D'-A'-3-5 is larger

Compare work during reversible (isentropic) expansion (B to C, vs. B' to C')
    Original: area B-C-10-8
    Modified: area B'-C'-9-7 is larger

Compare work during isothermal compression (C to D, vs. C' to D')
    Original: area C-D-6-10
    Modified: area C'-D'-5-9 is larger

Statements (b), (c), (d), and (e) are correct.

The <u>false</u> answer is (a) ●

THERMO 50

In the ideal heat pump system outlined, the expansion valve 4-1 performs the
process that is located on the T-S diagram between points

    (a)   A and B
    (b)   B and C
    (c)   C and D
    (d)   D and E
    (e)   E and A

The vapor compression reversed Rankine cycle is conducted counterclockwise on
both the schematic and T-S diagrams.
Numbers on the schematic and letters on the T-S diagram are related:
   1 = A, 2 = B, 3 = D, and 4 = E.
Process C-B-A occurs in the condenser between 2 and 1. The expansion process
A - E occurs between 1 - 4.

Answer is (e) ●

THERMO 51

Which air-standard power cycle do the P-V and T-S diagrams on the next page
represent?

    (a)   Otto cycle
    (b)   Reheat cycle
    (c)   Carnot cycle
    (d)   Rankine cycle
    (e)   Brayton cycle

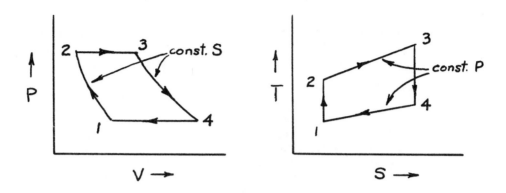

The Brayton cycle is applied to the simple open-cycle gas turbine wherein intake air is compressed (1-2), combustion supplies thermal energy (2-3), and combustion products expand and drive the turbine (3-4), and exhaust at 4. ●

Answer is (e)

THERMO 52

Equilibrium conditions exist in all **except** which of the following:

(a) in reversible processes.

(b) in processes where driving forces are infinitesimals.

(c) along ideal frictionless, nondissipative paths where forward and reverse processes occur at equal rates.

(d) in a steady state flow process.

(e) where nothing can occur without an effect on the system's surroundings.

All except (d) demonstrate equilibrium. Steady state or time invarient flow processes may be conducted far from equilibrium conditions. ●

The **false** answer is (d)

THERMO 53

Data in the table below describe two states of a working fluid that exist at two locations in a piece of hardware.

	P psia	v ft^3/lb	T °F	h BTU/lb	s BTU/(lb°F)
State 1	25.	0.011	20.	19.2	0.0424
State 2	125.	0.823	180.	203.7	0.3649

Which of the following statements about the path from state 1 to 2 is **false?**

(a) the path results in an expansion.

(b) the path determines the amount of work done.

(c) the path is indeterminate from these data.

(d) the path requires that energy be added in the process.

(e) the path is reversible adiabatic.

The large volume and entropy changes indicate change from a condensed to vapor phase. Temperature, pressure, and enthalpy increases require an energy input.

The path from 1 to 2 is indeterminate since no information on intermediate states is given. Work is always path dependent. Entropy increase means the process cannot be reversible adiabatic (isentropic).

The <u>false</u> answer is (e) ●

THERMO 54

Name the process that has no heat transfer.

(a) Isentropic
(b) Isothermal
(c) Quasistatic
(d) Reversible
(e) Polytropic

An <u>isentropic</u> process is reversible adiabatic. An <u>adiabatic</u> process has no heat exchange with surroundings. An <u>isothermal</u> process is conducted at constant temperature. A <u>quasistatic</u> (almost static) process departs only infinitesimally from an equilibrium state. A <u>reversible</u> process can have its initial state restored without any change (energy gain or loss) taking place in the surroundings. A <u>polytropic</u> process is conducted with changes in temperature, pressure, volume, and entropy; it follows the relationship PV^n = constant, where

$$n \neq \frac{C_p}{C_v}.$$

Answer is (a) ●

THERMO 55

In an open or flow process (across a fixed control volume) in the absence of PE and KE changes, which of the following represents the shaft work done during a polytropic process?

(a) $W_s = \dfrac{k}{1 - k}(P_2V_2 - P_1V_1) = \dfrac{k}{1 - k}mR(T_2 - T_1) = \dfrac{k}{1 - k}mRT_1\left[\left(\dfrac{P_2}{P_1}\right)^{\frac{k-1}{k}} - 1\right]$

(b) $W_s = P_1V_1\ln(\dfrac{P_1}{P_2}) = P_1V_1\ln(\dfrac{V_2}{V_1}) = mRT\ \ln(\dfrac{V_2}{V_1}) = mRT\ \ln(\dfrac{P_1}{P_2})$

(c) $W_s = \dfrac{n}{1 - n}(P_2V_2 - P_1V_1) = \dfrac{n}{1 - n}mR(T_2 - T_1) = \dfrac{n}{1 - n}mRT_1\left[\left(\dfrac{P_2}{P_1}\right)^{\frac{n-1}{n}} - 1\right]$

(d) $W_s = 0$

(e) $W_s = -V(P_2 - P_1)$

Shaft work, W_s, for a flow process $= -\int VdP$. The above equations are valid for ideal gases in the processes listed below:

(a) for constant entropy (isentropic, or reversible adiabatic) process.

(b) for isothermal process (at constant T, U and H are also constant for ideal gases).

(c) for polytropic process (where P, V, T, and S all change).

(d) for constant pressure (isobaric) process.

(e) for constant volume (isochoric) process.

The equations may be modified by manipulation with the ideal gas law
$PV = mRT$, $P_1V_1 = mRT_1$, $P_2V_2 = mRT_2$ in all cases. The relations that follow

also may be substituted, using n or k as appropriate, for polytropic and isentropic processes:

$$PV^n = P_1V_1^n = P_2V_2^n = \text{constant}, \qquad \left(\frac{P_2}{P_1}\right)^{\frac{n-1}{n}} = \frac{T_2}{T_1} = \left(\frac{V_1}{V_2}\right)^{\frac{n-1}{n}}$$

and $\dfrac{P_2}{P_1} = \left(\dfrac{V_1}{V_2}\right)^n$. The correct answer to the question is (c). All the rest are valid when applied to the proper <u>flow</u> process. ●

THERMO 56

In a closed system (with a moving boundary) which of the following represents work done during an isothermal process?

(a) $W = P(V_2 - V_1)$

(b) $W = 0$

(c) $W = P_1V_1 \ln(\frac{P_1}{P_2}) = P_1V_1 \ln(\frac{V_2}{V_1}) = mRT \ln(\frac{V_2}{V_1}) = mRT \ln(\frac{P_1}{P_2})$

(d) $W = \dfrac{P_2V_2 - P_1V_1}{1 - k} = \dfrac{mR(T_2 - T_1)}{1 - k}$

(e) $W = \dfrac{P_2V_2 - P_1V_1}{1 - n} = \dfrac{mR(T_2 - T_1)}{1 - n}$

Work for a closed system (piston-cylinder type, non-repetitious) $W = \int PdV$.

The above equations are valid for ideal gases in the processes listed below:
  (a) constant pressure
  (b) constant volume
  (c) isothermal process
  (d) isentropic process
  (e) polytropic process
Note that all modifications by manipulation noted in the previous question may be applied to this set of equations.

The Answer to the question is (c) ●

THERMO 57

Work of a polytropic (n = 1.21) compression of air ($\frac{C_p}{C_v}$ = 1.40) in a system with moving boundary from P_1 = 15. psia, V_1 = 1.0 ft^3 to P_2 = 150. psia, V_2 = 0.15 ft^3 is:

  (a)   35.5 ft.lb.
  (b)   324 ft.lb.
  (c)   1080 ft.lb.
  (d)   2700 ft.lb.
  (e)   5150 ft.lb.

Work of a closed system (moving boundary) polytropic process for an ideal gas is:

$$W = \frac{P_2V_2 - P_1V_1}{1 - n} = \frac{[150(0.15) - 15(1.0)]\,144}{1 - 1.21} = -5143. \text{ ft.lb.,} \; ●$$

work is done on gas.

Answer is (e)

THERMO 58

Isentropic compression of 1 ft^3 of air, $\dfrac{C_p}{C_v} = 1.40$, at 20 psia to a pressure of 100 psia gives a final volume of:

(a) 0.16 ft^3
(b) 0.20 ft^3
(c) 0.32 ft^3
(d) 0.40 ft^3
(e) 0.56 ft^3

An isentropic process for an ideal gas follows the path:

$$P V^k = P_1 V_1^{\,k} = P_2 V_2^{\,k} = \text{constant, where } k = \frac{C_p}{C_v}$$

$$20(1)^{1.4} = 100(V_2)^{1.4} \qquad V_2^{\,1.4} = 0.20 \qquad \therefore V_2 = 0.317 \text{ ft}^3 \; \bullet$$

Answer is (c)

THERMO 59

Steam enters a turbine at 300 psia, 700°F and exhausts to a low pressure process steam line at 30 psia. The turbine delivers 1000 Kw of power at an engine efficiency of 72%. Calculate the mass flow rate of steam in lb/hr.

Engine efficiency is the ratio of output to the available work corresponding to an isentropic expansion.

1 Kwh = 3413 BTU

Calculation basis: 1 hour

1000 Kwh at 72% efficiency is equivalent to 1389 Kwh, or

$W_s = 4.74 \times 10^6$ BTU from an isentropic expansion.

Incoming steam properties from superheated tables:

$$h_1 = 1368.3 \text{ BTU/lb} \qquad\qquad s_1 = 1.6751 \text{ BTU/(lb°F)}$$

Exhaust steam for isentropic expansion has $s_2 = 1.6751$ BTU/(lb°F) at 30 psia.

Consult saturated steam tables at 30 psia finding:

T, °F	h_f	h_{fg}	h_g	s_f	s_{fg}	s_g
250.34	218.83	945.2	1164.0	0.3680	1.3312	1.6992

Linearly interpolate entropy to 1.6751 and determine enthalpy on the basis of equal quality.

$s_2 - s_f = 1.6751 - 0.3680 = 1.2891$

s_{fg} for 100% quality $= 1.3312$ $\qquad X = \dfrac{1.2891}{1.3312} = 0.9684$

h_{fg} for 100% quality $= 945.2$

h_2 at .9684 quality $= 218.83 + 0.9684(945.2) = 1134.2$ BTU/lb.

Let $\dot{m}_1$ be steam flow rate in lb/hr.

$\dot{m}_1(h_1 - h_2) = W_s$

$\dot{m}_1(1368.3 - 1134.2) = 4.74 \times 10^6$ BTU

$$\dot{m} = \frac{4.74 \times 10^6}{234.1} = 20,200 \text{ lb/hr.} \;\bullet$$

THERMO 60

An ideal gas at a pressure of 500 psia and a temperature of 75°F is contained in a cylinder with a volume of 700 cubic feet. A certain amount of the gas is released so that the pressure in the cylinder drops to 250 psia. The expansion of the gas is isentropic. The heat capacity ratio is 1.40 and the gas constant is 53.3 ft lb_f/lb_m°R.

What is the weight of the gas remaining in the cylinder?

〜〜〜〜〜〜〜

Given: $k = \dfrac{C_p}{C_v} = 1.40$ $\qquad\qquad R = 53.3 \dfrac{\text{ft } lb_f}{lb_m \text{ °R}}$

$P_1 = 500$ psia $\qquad\qquad\qquad P_2 = 250$ psia

$V_1 = 700$ cu ft $\qquad\qquad\quad V_2 = 700$ cu ft

$T_1 = 75°F + 460 = 535°R \qquad \begin{array}{l} T_2 = ? \\ w_2 = ? \end{array}$

Basis:

Ideal gas law $P_1 V_1 = wRT_1$ and the basic equations for reversible adiabatic (isentropic) expansion

$$\frac{T_2}{T_1} = \left(\frac{V_1}{V_2}\right)^{k-1} \qquad \text{and} \qquad \frac{T_2}{T_1} = \left(\frac{P_2}{P_1}\right)^{\frac{k-1}{k}}$$

The gas remaining in the tank cools as it expands

$$T_2 = T_1 \left(\frac{P_2}{P_1}\right)^{\frac{k-1}{k}} = 535\left(\frac{250}{500}\right)^{\frac{1.4-1}{1.4}} = 535\left(\frac{1}{2}\right)^{0.2857} = 439°R$$

$P_2 V_2 = w_2 R T_2$ $\qquad$ Check dimensions to use the proper conversion factor.

$$250 \frac{lb_f}{in^2} \times \frac{144 \text{ in}^2}{ft^2} \times 700 \text{ ft}^3 = w_2 \; lb_m \frac{53.3 \text{ ft } lb_f}{lb_m \text{ °R}} \times 439°R$$

$$w_2 = \frac{(250)(144)(700)}{(53.3)(439)} = 1077 \; lb_m \;\bullet$$

THERMO 61

Determine the theoretical horsepower required for the isothermal compression of 800 ft^3/min of air from 14.7 to 120 psia.

At this volume flow rate, the process is a flow process. Work of <u>isothermal compression</u> of an ideal gas is numerically the same in a steady flow process as in a closed system.

PV = constant = P_1V_1 = P_2V_2 = RT

closed system: $W = \int_{V_1}^{V_2} PdV = P_1V_1 \ln\frac{V_2}{V_1} = P_1V_1 \ln\frac{P_1}{P_2} = mRT \ln\frac{V_2}{V_1} = mRT \ln\frac{P_1}{P_2}$

flow system: $W_s = -\int_{P_1}^{P_2} VdP = -P_1V_1 \ln\frac{P_2}{P_1} = P_1V_1 \ln\frac{P_1}{P_2} = P_1V_1 \ln\frac{V_2}{V_1}$

$= mRT \ln\frac{V_2}{V_1} = mRT \ln\frac{P_1}{P_2}$

Basis of calculation: 1 minute

$W_s = P_1V_1 \ln\frac{P_1}{P_2} = 14.7(144)(800) \ln(\frac{14.7}{120}) = -3,560,000$ ft.lb/min, work in

$W_s = \frac{3,560,000 \text{ ft.lb/min}}{33,000 \text{ ft.lb/HP-min}} = 108.$ HP ●

THERMO 62

Enthalpy of an ideal gas is a function only of:
      (a)   internal energy
      (b)   entropy
      (c)   the product of pressure and specific volume
      (d)   pressure
      (e)   temperature

$H = U + PV$, $PV = RT$.   Enthalpy and internal energy of <u>ideal</u> gases are a function only of temperature. $dh = C_p dT$ and $du = C_v dT$.

Answer is (e) ●

THERMO 63

Which of the following statements is <u>false</u> concerning the deviations of real gases from ideal gas behavior?

    (a) Molecular attraction interactions are compensated for in the ideal gas law.

    (b) Deviations from ideal gas behavior are large near the saturation curve.

    (c) Deviations from ideal gas behavior become significant at pressures above the critical point.

    (d) Molecular volume becomes significant as specific volume is decreased.

    (e) Compressibility factor Z is used to modify the ideal gas equation of state to fit real gas behavior.

All statements except (a) are true. The ideal gas law does not consider volume of the molecules or any interaction other than elastic collision. ●

The <u>false</u> statement is (a)

THERMO 64

There are 3 lbs. of air in a rigid container at 25 psia and $100°F$. Given the gas constant for air is 53.35
(a) Determine the volume of the container
(b) If the temperature is raised to $180°F$, what is the resulting absolute pressure?

$PV = mRT$

$$\frac{1b}{ft^2} \cdot ft^3 = 1b_m \cdot \frac{ft - 1b}{1b_m °R} \cdot °R$$

$$P = 25 \text{ psia} \times 144 = 3600 \text{ 1b/ft}^2$$
$$R = 53.35 \text{ ft-1b/1b}_m \cdot °R$$
$$T_1 = 100°F + 460 = 560°R$$
$$T_2 = 180°F + 460 = 640°R$$

(a) $PV = mRT$

$$3600(V) = 3(53.35)(560) \qquad V = \frac{3(53.35)(560)}{3600} = 24.9 \text{ ft}^3 ●$$

(b) $\dfrac{P_1 V_1}{T_1} = \dfrac{P_2 V_2}{T_2}$ \qquad For constant volume

$$\frac{P_1}{T_1} = \frac{P_2}{T_2}$$

$$P_2 = \frac{P_1 T_2}{T_1} = \frac{25(640)}{560} = 28.6 \text{ psia} ●$$

THERMO 65

A mixture at 14.7 psia and $68°F$ that is 30% weight CO_2 (m wt = 44) and 70% weight N_2 (m wt = 28) has a partial pressure of CO_2 in psia that is nearest to:

(a) 2.14
(b) 3.15
(c) 6.83
(d) 7.86
(e) 11.55

Basis of calculation: 1 lb. mixed gases

(1) Calculate weight of each component, number of moles of each present.

(2) Compute mole fraction of each, and proportion total pressure according to mole fraction.

Component	Weight, lb.	Number of lb. moles	Mole Fraction	Partial Pressure psia
CO_2	0.30	$\dfrac{0.30}{44} = .00683$	$\dfrac{.00683}{.03183} = 0.214$	3.15
N_2	0.70	$\dfrac{0.70}{28} = 0.0250$	$\dfrac{0.0250}{0.03183} = 0.786$	11.55
Total	1.00	0.03183	1.000	14.70

Since mole fraction of a gas = volume fraction,

Composition of the mixture is 21.4% vol. CO_2 and 78.6% vol. N_2.

From the table, the correct partial pressure of CO_2 is 3.15 psia. ●

Answer is (b)

THERMO 66

Given that molar C_p of CO_2 is 8.92 BTU/(lb.mole°R), and molar C_p of N_2 is 6.95 BTU/(lb.mole°R), the calculated C_p per pound of mixture containing 25% vol. CO_2 and 75% vol. N_2 is:

(a) 0.23 BTU/(lb.°R)

(b) 2.23 BTU/(lb.°R)

(c) 5.21 BTU/(lb.°R)

(d) 7.44 BTU/(lb.°R)

(e) none of these

Note that theoretical molar heat capacities of ideal gases at zero pressure are for monotomic: $C_v = 3$, $C_p = 5$. BTU/(lb.mole°R)

     diatomic: $C_v = 5$, $C_p = 7$.

     triatomic: $C_v = 7$, $C_p = 9$.

Basis of calculation: 1 lb. mole of mixture

Component	Mol. Wt.	% vol.	Mole fraction	Weight of each component	Molar C_p contribution
CO_2	44	25.	0.25	11.0	0.25(8.92) = 2.23
N_2	28	75.	0.75	21.0	0.75(6.95) = 5.21
Total		100.	1.00	32.0	7.44

Average Mol.
Weight

One mole of mixture weighs 32.0 lb. and has a Molar C_p of 7.44 BTU/(lb.mole°R)

The mixture has a specific C_p of $\frac{7.44}{32.0}$ = 0.232 BTU/(lb.°R) ●

Answer is (a)

THERMO 67

Dry air has an average molecular weight of 28.9, consisting of 21 mole-% O_2, 78 mole-% N_2 and 1 mole-% (Argon, traces of CO_2). Its calculated wt.% O_2 is nearest to:

(a) 21.0

(b) 22.4

(c) 23.2

(d) 24.6

(e) 28.0

Basis of calculation: 1 lb. mole dry air

Component	Mol. Wt.	Mole Fraction	Weight, lb.	% Weight
O_2	32.0	0.21	6.72	23.2
N_2	28.0	0.78	21.80	75.4
Ar	40.0	0.01	0.40	1.4
Total		1.00	28.92	100.0

Answer is (c) ●

THERMO 68

All of the following statements about wet bulb temperature are true EXCEPT:

    (a) wet bulb temperature equals adiabatic saturation temperature
    (b) wet bulb temperature lies numerically between dewpoint and dry bulb temperatures for unsaturated systems
    (c) wet bulb temperature equals both dry bulb and dewpoint temperature at 100% relative humidity
    (d) wet bulb temperature is the only temperature necessary to determine grains of water per lb. of dry air
    (e) wet bulb temperature is the lowest temperature attainable by evaporative cooling at ambient pressure

Wet bulb temperature, as commonly measured, and adiabatic saturation temperature are equal for the air-water system. For unsaturated systems, wet bulb temperature is approximately equal to dewpoint plus 1/3 of the difference between dry bulb and dew point temperatures. At saturation, wet bulb, dry bulb and dewpoint are equal. Wet bulb temperature alone is insufficient to determine state on a psychrometric chart, hence insufficient for determining specific humidity ratio (grains of water/lb. dry air) or any other property. Both wet and dry bulb temperatures are normally used to locate a point on a psychrometric chart. Wet bulb (adiabatic saturation) temperature is normally the lowest temperature obtained by evaporative cooling at ambient pressure.

Answer is (d) ●

THERMO 69

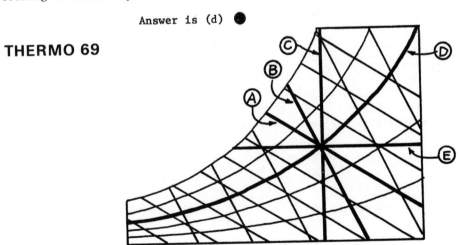

Line E of the usual psychrometric chart above represents a line of constant:

    (a)   relative humidity
    (b)   dew point or specific humidity
    (c)   enthalpy or wet bulb temperature
    (d)   specific volume
    (e)   dry bulb temperature

Sloped lines A are lines of constant wet bulb temperature and are almost lines of constant enthalpy.
Steeper lines B are lines of constant specific volume, cu ft wet air/lb dry air.
Vertical lines C are lines of constant dry bulb temperature.
Curved lines D are lines of constant relative humidity.
Horizontal lines E are lines of constant dew point, specific humidity and water vapor pressure.

Note: Psychrometric charts are only valid at one pressure, usually 14.7 psia, though they are usable with little error for barometric pressures between 29-31 " Hg.   Answer is (b) ●

THERMO 70

A small plastic bag is filled with moist air at 50% relative humidity, sealed and placed in an environmental chamber whose temperature and pressure may be independently varied. The relative humidity in the bag will be lowered under which of the following conditions?

(a) pressure is increased
(b) pressure is decreased
(c) temperature is decreased
(d) temperature is decreased and pressure is increased
(e) none of the above conditions

〰〰〰〰〰〰

Unless water condensation occurs at 100% R.H., the sealed bag contains a fixed mole fraction of water vapor in air.

$$\text{Relative Humidity} = \left\{ \frac{\text{mole fraction of water vapor in the mixture}}{\begin{array}{c}\text{mole fraction of water vapor in a saturated mixture}\\ \text{at the same temperature and pressure.}\end{array}} \right\}$$

Vapor pressure of water is temperature dependent only, and increases with temperature. The denominator term of the above R.H. definition is increased with elevated temperature and decreased with elevated pressure. Thus: pressure increase raises R.H., pressure decrease lowers R.H., temperature decrease raises R.H., while simultaneous temperature decrease and pressure increase both raise R.H. Note that a temperature increase lowers R.H., but that choice was not given in the problem.

Answer is (b)   ●

THERMO 71

Which of the following statements about stagnation properties is <u>least</u> correct?

(a) For a compressible fluid, KE change is converted to stagnation temperature rise, and pressure rise is isentropically calculated.

(b) Stagnation temperature is related to aerodynamic heating of leading edges of aircraft wings.

(c) For an incompressible fluid impacting a pitot tube, KE change may be converted to pressure rise.

(d) For gases at low velocity, the simplifying assumption of incompressibility yields stagnation pressures with only a small percentage error.

(e) In a fluid at Mach number greater than 1, an isentropic calculation of stagnation properties is valid.

〰〰〰〰〰〰

Whenever a flow is brought to rest isentropically, a stagnation pressure arises. At Mach ≻ 1 a shock wave forms in front of the body and dissipates energy by a process which is not isentropic. Stagnation pressures produced at the body are lower than anticipated.

Answer (e) is false ●

THERMO 72

Which of the following statements about Mach number (M) is <u>false</u>?

    (a)   Mach number is the ratio of velocity to sonic velocity.

    (b)   Mach 1 is the maximum attainable velocity in a nozzle throat.

    (c)   Supersonic velocities (M > 1) are achievable in the diffuser section of a rocket nozzle if expansion ratio is great enough.

    (d)   Mach angle α, or angle the shock front makes with the velocity vector of a moving source, and Mach number are related by $M\sin\alpha = 1$.

    (e)   Mach no. has the dimensions of velocity.

Mach number, being a ratio of velocities, is dimensionless. All other statements are correct.

Answer (e) is false ●

THERMO 73

Sonic velocity (Mach 1) at 50,000 feet altitude in the standard upper atmosphere (-67.6°F and 1.68 psia) is nearest to:

    (a)   880. ft/sec
    (b)   971. ft/sec
    (c)   995. ft/sec
    (d)   1064. ft/sec
    (e)   1117. ft/sec

Sonic velocity for ideal gases is temperature dependent only and may be calculated from

$$C = \sqrt{kg_c RT} \qquad \text{where} \quad C \text{ is sonic velocity in ft/sec.}$$

$$k = \frac{C_p}{C_v}$$

$$g_c = 32.17 \; lb_m \; ft/(lb_f \; sec^2)$$

$$R = \text{gas constant} \quad ft \; lb_f/(lb_m \, °R)$$

$$T = \text{temperature in } °R.$$

$$C = \sqrt{1.40(32.17)(53.34)(392.4)} \; = 971 \; ft/sec \; ●$$

Answer is (b)

THERMO 74

Air flows through a 1 ft diameter circular duct at 20 psia and 80°F. A manometer connects a pitot tube, located on the centerline of the duct, with a static pressure tap upstream and reads a differential pressure of 3 inches of water. Determine mass flow rate of the air in lb/sec.

(1) <u>Solution based on assumed incompressibility</u>

Since differential pressure is trivial, this is acceptable.
  (3" water = 0.25' water = 0.25(62.4) = 15.6 psf = 15.6/144 = 0.108 psi)

In the incompressible solution, KE is converted to an additional pressure energy indicated by the pitot tube, and velocity is determined.

Basis: 1 lb_m of fluid flowing.

$$\frac{v^2}{2g_c} = h = \frac{\Delta P}{\rho} \qquad \text{where} \qquad \begin{array}{l} h \text{ is additional head in feet} = ft\ lb_f/lb_m \\ \rho \text{ is density of the fluid flowing in } lb_m/ft^3 \\ \Delta P \text{ is differential pressure in } lb_f/ft^2 \\ V \text{ is velocity in ft/sec} \\ g_c \text{ is 32.17 } lb_m\ ft/(lb_f\ sec) \end{array}$$

For air, $R = 53.34\ \dfrac{ft\ lb_f}{lb_m\ {}^\circ R}$ and density is the reciprocal of specific volume,

  $v = \dfrac{1}{\rho}$.

$\rho = \dfrac{P}{RT} = \dfrac{20(144)}{53.34(540)} = 0.10\ lb/ft^3 =$ density of air in the duct

$\dfrac{v^2}{2(32.17)} = \dfrac{15.6}{0.10} = 156. = h = \dfrac{\Delta P}{\rho}$

This pressure energy (flow energy) is equivalent to a 156. foot (head) column of air at the T and P existing in the duct.

$V = \sqrt{2(32.17)(156)}\ = 100\ ft/sec.$

Duct cross-sectional area $= \dfrac{\pi}{4}\ d^2 = 0.785\ ft^2 = A$

Volume flow rate $= 0.785\ ft^2(100\ ft/sec) = 78.5\ cfs. = AV = q$

Since $\dot{m} = q\ \rho$, $\dot{m} = 78.5\ \dfrac{ft^3}{sec} \times 0.10\ \dfrac{lb}{ft^3} = 7.85\ lb_m/sec.$ ●

(2) <u>Solution based on compressible fluid</u>

KE is converted to heat, resulting in a stagnation temperature, T_{stag} and this has a corresponding isentropic pressure rise, ΔP to stagnation pressure, P_{stag}.

Basis: 1 lb_m of fluid flowing

$$\frac{v^2}{2g_c J} = C_p \Delta T \qquad\qquad \begin{array}{l} T_{stag} = T_1 + \Delta T = T_2 \\ P_{stag} = P_1 + \Delta P = P_2 \end{array}$$

For an isentropic compression:

$\left(\dfrac{P_2}{P_1}\right)^{\frac{k-1}{k}} = \dfrac{T_2}{T_1}$ where $k = \dfrac{C_p}{C_v}$ For air: $\begin{array}{l} C_p = 0.240\ BTU/lb\,{}^\circ R \\ C_v = 0.171\ BTU/lb\,{}^\circ R \\ k = 1.40 \end{array}$

  (k is often referred to as γ)

These ratios may be handled using any consistent <u>absolute</u> units.

$$\left(\frac{20 + 0.108}{20}\right)^{\frac{1.40-1}{1.40}} = \frac{T_2}{540} \qquad (1.00540)^{0.2857} = \frac{T_2}{540}$$

$$1.00154 = \frac{T_2}{540} \qquad \therefore T_2 = 540.832°R \quad or \quad \Delta T = 0.832°R$$

To avoid inaccuracy, it is necessary to use as many digits as possible.

$$V^2 = 2g_c J C_p \Delta T \quad where \quad J = 778 \text{ ft lb/BTU}$$

$$V = \sqrt{2(32.17)(778)(0.240)(0.832)} = 100. \text{ ft/sec.}$$

Note that this result agrees with the incompressible easier case where velocities are low and T_{stag} and P_{stag} are not much different from the duct temperature and pressure. The compressible solution is the only valid one at very high velocities.

Summary: $V = 100$ ft/sec $\therefore \dot{m} = 7.85$ lb/sec ●

THERMO 75

During complete stoichiometric combustion of 1 lb. mole of methane (CH_4) with air, the number of lb. moles of nitrogen and other inerts that pass through the combustion zone is nearest to:

(a) 0.79
(b) 3.9
(c) 5.6
(d) 7.5
(e) 16.1

$$CH_4 + 2O_2 \longrightarrow CO_2 + 2H_2O$$

1 lb. mole of methane requires 2 lb. moles of O_2 for complete combustion.

Air is 21% vol. O_2 and 79% vol. (N_2 + inerts).

2 lb. moles of O_2 is supplied by $\frac{2}{0.21} = 9.52$ lb. moles of air, which contains 7.52 lb. moles of (N_2 + inerts). ●

Answer is (d)

THERMO 76

During combustion of hydrocarbon fuels with <u>less</u> than stoichiometric air, the products depend on the amount of air supplied and could be all of the following <u>except</u>:

(a) Unburned fuel, C, CO_2 and H_2O
(b) Unburned fuel, C, CO and H_2O
(c) C, CO, CO_2 and H_2O
(d) C, CO and H_2O
(e) CO, CO_2 and H_2O

Combustion of hydrocarbon fuels occurs stepwise in the following sequence:
first: all hydrogen to H_2O, leaving carbon

next: all carbon to CO

last: all CO to CO_2

This is summarized in the table below:

Amount of Air	Products from Combustion Zone				
	Unburned Fuel	Carbon	CO	CO_2	H_2O
excess air				X	X
stoichiometric				X	X
slightly fuel rich			X		X
less than stoichiometric air		X	X		X
	X	X	X		X
very fuel rich	X	X			X

As a practical matter it is difficult to obtain a CO-free stack gas due to slow kinetics of CO combustion to CO_2, and due to equilibrium in the water gas shift reaction.

Choice (a) in the problem does **not** occur ●

THERMO 77

(a) Write the combustion equation for a heavy oil whose composition averages $C_{17}H_{36}$ with complete combustion in theoretical air.

(b) Determine the pounds of theoretical air per pound of $C_{17}H_{36}$ fuel.

~~~~~~~~~~~

(a) During stoichiometric combustion all hydrogen is burned to water and carbon to CO_2.

$$C_{17}H_{36} + 26O_2 \longrightarrow 17CO_2 + 18H_2O \quad ●$$

From the calculation below, 97.8 (N_2 + inerts) pass through unreacted.

(b) Basis of calculation: 1 lb. mole of $C_{17}H_{36}$

26 lb. moles of O_2 are required. Air is 21% vol. (21 mole%) O_2 and 79% vol. (79 mole%) N_2 + inerts.

$$\frac{26 \text{ lb. moles } O_2}{21} = \frac{X \text{ lb. moles } (N_2 + \text{inerts})}{79} = \frac{Y \text{ lb. moles air}}{100}$$

X = 97.8 lb. moles (N_2 + inerts) that pass through the combustion zone

and Y = 123.8 lb. moles air required.

Since mol. wt. of $C_{17}H_{36}$ = 17(12) + 1(36) = 240

and the mol. wt. of air is 28.95

$$\frac{\text{lb air}}{\text{lb fuel}} = \frac{123.8(28.95)}{1(240)} = 14.9 \text{ air/fuel wt. ratio} \quad ●$$

THERMO 78

Gasoline has a heat of combustion of 21,000 BTU/lb and a specific weight of 6.17 lb/gal. Food has an average heat of combustion of roughly 6000 BTU/lb, and the average human's rate of energy comsumption is approximately 12,000 BTU/day. The automobile's average rate of fuel consumption in city driving is 3 gallons per hour.

Given that the equation for the combustion of both gasoline and food is
$$Fuel + Oxygen = Carbon\ Dioxide + Water,$$
and given that the chemical formulas for gasoline and food are C_8H_{18} and $C_6H_{12}O_6$, respectively, compute the ratio of the amount of air used per hour by the automobile to the amount used by an average human.

Basis of calculation: 1 hour

(1) <u>Automobile</u>
Stoichiometric equation: $\qquad C_8H_{18} + 12.5\ O_2 \longrightarrow 8CO_2 + 9H_2O$

1 lb. mole of fuel, mol. wt. = 8(12) + 18(1) = 114 lb. fuel, requires 12.5 lbmoles of oxygen

Fuel consumption = 3 gal/hr x 6.17 lb/gal = 18.51 lb/hr.

Oxygen requirement is obtained by ratio:

$$\frac{18.51}{114} = \frac{lbmoles\ oxygen}{12.5}$$

Oxygen = 2.030 lb moles/hr.

(2) <u>Human</u>
Stoichiometric equation: $\qquad C_6H_{12}O_6 + 6O_2 \longrightarrow 6CO_2 + 6H_2O$

1 lb. mole of food, mol. wt. = 6(12) + 12(1) + 6(16) = 180 lb. food, requires 6 lb. moles of oxygen.

Food consumption = $\dfrac{12,000\ BTU/day}{6,000\ BTU/lb.} \times \dfrac{1\ day}{24\ hr}$ = 0.0833 lb/hr.

Oxygen requirement is obtained by ratio:

$$\frac{0.0833}{180} = \frac{lbmoles\ oxygen}{6}$$

Oxygen = 0.00278 lb moles/hr.

(3) <u>Comparison</u>

$$\frac{Auto\ Air}{Human\ Air} = \frac{Auto\ Oxygen}{Human\ Oxygen} = \frac{2.030}{0.00278} = 731. \quad \bullet$$

THERMO 79

Propane (C_3H_8) is burned in a heater with 20% excess air at ambient conditions (68°F, 14.7 psia, 30% relative humidity). Assume complete combustion to water and carbon dioxide.

Per pound mole of propane, calculate:

      (a)   volume of stack gases cooled to 800°F.

      (b)   composition of stack gases (vol.%).

      (c)   dew point of stack gases.

Calculation basis: 1 lb. mole of C_3H_8

(1) <u>Stoichiometric equation</u>: $C_3H_8 + 5O_2 \longrightarrow 3CO_2 + 4H_2O$

5 lb. moles O_2 are required for complete combustion. At 20% excess air, 6 lb. moles enter combustion zone accompanied by some moisture and much nitrogen plus inerts.

(2) <u>Calculation of moisture and N_2 plus inerts carried by intake air</u>

From the saturated steam table, water vapor pressure at 68°F is 0.339 psia; this corresponds to 100% relative humidity. At 30% R.H. the water vapor pressure is (0.30)(0.339) = 0.102 psia.

Mole fraction of water vapor is $\frac{0.102}{14.7}$ = 0.00695, or 0.695% and mole fraction of dry air is 0.993.

Alternately, from a psychrometric chart, the water contained in intake air is 31. grains/(lb. dry air). At 7000 grains/lb., this is 0.0044 lb. H_2O/(lb. dry air).

$$\text{Mole fraction of water vapor} = \frac{\text{moles water vapor}}{\text{moles dry air} + \text{moles water vapor}}$$

$$= \frac{\frac{0.0044}{18}}{\frac{1}{28.9} + \frac{0.0044}{18}} = 0.070$$

Mole fraction of dry air = 1 - 0.070 = 0.993

6 lb. moles O_2 are carried by $\frac{100}{21}(6)$ = 28.6 lb. moles dry air or by $\frac{28.6}{0.993}$ = 28.8 lb. moles moist air containing 0.2 lb. moles H_2O.

Nitrogen + inerts carried = 28.6 - 6.0 = 22.6 lb. moles.

(3) <u>Summary: lb. moles entering and leaving combustion zone</u>

	C_3H_8	O_2	N_2 plus inerts	H_2O	CO_2	Total
Entering	1.0	6.0	22.6	0.2	0	29.8
Leaving	0	1.0	22.6	0.2+4.0 = 4.2	3.0	30.8

Note from the stoichiometric equation that 6 moles of reactants form 7 moles of products - justifying change in moles in and out above.

(4) <u>Computation of stack gas volume</u>

Use ideal gas law $PV = nRT$, where $R = \frac{1545 \text{ ft lb}_f}{\text{lb}_m \text{ mole °R}}$, to calculate volume of 30.8 moles stack gases at 800 F (1260°R) and 14.7 psia.

$14.7(144)V = 30.8(1545(1260)$ $\therefore V = 28,300 \text{ ft}^3$

Alternate calculation based on 1 lb.mole = 359 ft^3 at 32°F, 14.7 psia:

$$30.8(359) \frac{1260°R}{492°R} = 28,400 \text{ ft}^3 \ \bullet$$

(5) <u>Stack gas composition</u>   (vol.% = mole %)

Component	Excess O_2	N_2 plus inerts	CO_2	H_2O	Total
lb. moles	1.0	22.6	3.0	4.2	30.8
vol. %	3.25	73.35	9.75	13.65	100.00 ●

(6) <u>Dew point of stack gas</u>

Determine temperature where vapor pressure of water equals partial pressure of water in stack gas.

From above table, vol. % H_2O = 13.65   mole fraction = 0.1365

Water partial pressure in stack gas = (14.7 psia)(0.1365) = 2.01 psia.

Consult saturated steam table for vapor pressure of water at various temperatures.   Find:

> 120°F   1.692 psia
> 130°F   2.221 psia

Although vapor pressure does not vary linearly with temperature, linear interpolation is adequate over a short temperature range and is as accurate as the problem input data justify. Conclude that dew point = 126°F. ●

> > Summary of results
> > > (a) Stack gas volume = 28,300 cf/lb.mole C_3H_8 ●
> > > (b) Stack gas composition, vol. % (see table above) ●
> > > (c) Stack gas dew point = 126°F ●

THERMO 80

Theoretically, how many pounds of air are required for complete combustion of 10 pounds of ethane gas (C_2H_6), and how many pounds of water vapor are produced in the products of combustion?
Assume the air contains 23% oxygen by weight.

Write and balance the combustion equation

$$2C_2H_6 + 7O_2 \longrightarrow 4CO_2 + 6H_2O$$

The molecular weights must be calculated

> C = 12   H = 1   O = 16

Therefore C_2H_6 = 30 O_2 = 32 CO_2 = 44 H_2O = 18

Thus 2 moles of C_2H_6 (2 x 30 lbs) plus 7 moles of O_2 (7 x 32 lbs)

> yield 4 moles of CO_2 (4 x 44 lbs) plus 6 moles of H_2O (6 x 18 lbs).

Therefore 10 lbs of C_2H_6 require

$$\frac{10}{60}(7 \times 32) = 37.33 \text{ lbs of oxygen}$$

Since air is 23% oxygen by weight,

> pounds of air required = $\frac{37.33}{0.23}$ = 162 pounds ●

10 lbs of C_2H_6 will produce

$$\frac{10}{60}(6 \times 18) = 18 \text{ lbs of water vapor} \; ●$$

THERMO 81

If the heat of combustion of kerosene ($C_{10}H_{22}$) is 1625 K cal/mole and kerosene costs 25 cents per gallon, what is the cost of 25 K cal obtainable by burning kerosene? Assume complete combustion of kerosene, the density of kerosene is 0.80 grams per c.c., and that one fluid ounce is equal to 30 c.c.

Atomic Weights   O - 16   C - 12   H - 1

$\sim\sim\sim\sim\sim\sim$

mole = gram-molecular weight

number of moles burned = $\dfrac{25}{1625}$ = $\dfrac{1}{65}$ mole kerosene

$C_{10}H_{22}$ has a molecular weight of 10(12) + 22(1) = 142

kerosene burned = $\dfrac{1}{65}$ x 142 = 2.185 grams

kerosene cost = $\dfrac{2.185 \text{ gms}}{0.8 \frac{\text{gms}}{\text{cc}}}$ x $\dfrac{1}{30 \frac{\text{cc}}{\text{fl oz}}}$ x $\dfrac{1}{32 \frac{\text{fl oz}}{\text{quart}}}$ x $\dfrac{1}{4 \frac{\text{qt}}{\text{gal}}}$ x 25 $\dfrac{\text{cents}}{\text{gal}}$

= 0.0178 cents ●

THERMO 82

Given the following data for propane:

Formula C_3H_8
Weight (liquid)   4.24 lb/gal
Specific Volume   8.55 ft^3 gas/lb

REQUIRED: Compute the following:

(a) lbs O_2 per lb of gas for complete combustion

(b) lbs CO_2 produced per lb of gas burned

(c) lbs H_2O produced per lb of gas burned

$\sim\sim\sim\sim\sim\sim\sim$

First, write the balanced chemical equation.

$$C_3H_8 + 5O_2 \longrightarrow 3CO_2 + 4H_2O$$

Knowing the atomic weights, compute the molecular weights.

        44   32   44   18   (C = 12, O = 16, H = 1)

So, 44 pounds of C_3H_8 combine with 5 x 32 = 160 pounds of oxygen to form 3 x 44 = 132 pounds of CO_2 plus 4 x 18 = 72 pounds of H_2O.

(a) $\dfrac{160}{44}$ = 3.64 pounds of O_2 per pound of C_3H_8 ●

(b) $\dfrac{132}{44}$ = 3.0 pounds of CO_2 per pound of C_3H_8 ●

(c) $\dfrac{72}{44}$ = 1.64 pounds of H_2O per pound of C_3H_8 ●

THERMO 83

The temperature difference between the two sides of a solid rectangular slab of area A and thickness L as shown below is ΔT. The heat transferred through the slab by conduction in time t is proportional to:

(a) $AL\Delta T\, t$

(b) $AL\, \dfrac{\Delta T}{t}$

(c) $AL\, \dfrac{t}{\Delta T}$

(d) $\dfrac{A\Delta T\, t}{L}$

(e) $\dfrac{A(\Delta T)^2 t}{L}$

Heat transferred through slab by conduction in time t is:

$$Q = kA\Delta T \quad \text{where} \quad k = \frac{BTU\ ft}{hr\ ft^2\ {}^{\circ}F}$$

$$Q = k\left|\frac{BTU\ ft}{hr\ ft^2\ {}^{\circ}F}\right| A\, \frac{(ft^2)(\Delta T)^{\circ}F\ t\ hrs}{L\ ft} = BTU = \frac{k(A\Delta T\, t)}{L}$$

k is the coefficient of thermal conductivity of the material, hence heat transfer in a given material is proportional to the other variables.

Answer is (d) ●

THERMO 84

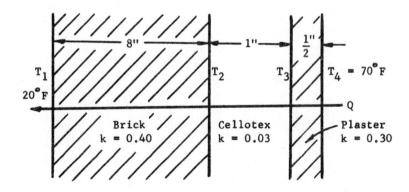

Given the composite wall shown with steady state outer temperature $T_1 = 20^{\circ}F$ and inner temperature $T_4 = 70^{\circ}F$, which of the following is nearest to T_3 in ${}^{\circ}F$?

(a) 27.
(b) 38.
(c) 46.
(d) 58.
(e) 69.

At steady state the same Q flows across each material and temperatures descend in direct relation to thermal resistance (reciprocal of conductivity).

$$\text{Resistance of brick} = \frac{x}{k} = \frac{0.667 \text{ ft}}{0.40 \text{ BTU}/(\text{ft}^2{}^\circ\text{F}/\text{ft})} = 1.67 \text{ ft}^2{}^\circ\text{F}/(\text{BTU})$$

$$\text{Resistance of Cellotex} = \frac{x}{k} = \frac{0.083 \text{ ft}}{0.03} = 2.77$$

$$\text{Resistance of plaster} = \frac{x}{k} = \frac{0.042 \text{ ft}}{0.30} = \underline{0.14}$$

$$\text{Total Resistance} = 4.58 \text{ ft}^2{}^\circ\text{F}/(\text{BTU})$$

$$Q = \frac{\Delta T \text{overall}}{\text{total resistance}} = \frac{\Delta T \text{ for plaster}}{\frac{x}{k} \text{ for plaster}} = \cdots$$

$$Q = \frac{50}{4.58} = \frac{T_4 - T_3}{0.14} = \frac{T_3 - T_2}{2.77} = \frac{T_2 - T_1}{1.67} = 10.85 \text{ BTU/hr.}$$

$T_4 - T_3 = 1.5^\circ\text{F}$ Since $T_4 = 70.^\circ\text{F}$, $T_3 = 68.5^\circ\text{F}$ ●

$T_3 - T_2 = 30.3^\circ\text{F}$ Since $T_3 = 68.5^\circ\text{F}$, $T_2 = 38.2^\circ\text{F}$

$T_2 - T_1 = 18.2^\circ\text{F}$ Since $T_2 = 38.2^\circ\text{F}$, $T_1 = 20^\circ\text{F}$ (in agreement with given data)

Answer is (e)

THERMO 85

The rate of heat transfer through a given section of a uniform wall for a given temperature difference is:

(a) Directly proportional to the thermal conductivity and to the thickness of the wall.

(b) Inversely proportional to the thermal conductivity and directly proportional to the thickness of the wall.

(c) Directly proportional to the thermal conductivity and inversely proportional to the thickness of the wall.

(d) Inversely proportional to the thermal conductivity and to the thickness of the wall.

(e) Independent of the thickness of the wall.

$$Q = \frac{k \, A(\Delta T)}{L}$$

where   k = conductivity
A = surface area
ΔT = temperature difference
L = wall thickness

Answer is (c) ●

THERMO 86

The heat loss per hour through 1 sq ft of furnace wall 18" thick
is 520 Btu. The inside wall temperature is 1900°F, and its average
thermal conductivity is 0.61 Btu/hr ft °F.

The outside surface temperature of the wall is nearest to

(a)   100°F
(b)   300°F
(c)   600°F
(d)   1000°F
(e)   1900°F

$$q = k \frac{A}{L}(t_1 - t_2) \quad \text{where} \quad t_1 = 1900°F \quad t_2 = \text{outside temp.}$$

$$k = 0.61 \text{ Btu/hr ft °F}$$

$$\frac{q}{A} = 520 \text{ Btu/hr}$$

$$L = 1.5 \text{ ft}$$

Solving for t_2

$$t_2 = -\frac{q}{A}\frac{L}{k} + t_1 = -520\frac{1.5}{0.61} + 1900 = -1279 + 1900 = 621°F \; \bullet$$

Answer is (c)

THERMO 87

Heat is transferred by conduction from left to right through the composite
wall shown in the drawing below:

Assume the three materials are in good thermal contact and no significant film
coefficients exist at any of the interfaces. Determine which of the following
numbers most nearly represents the overall coefficient U in BTU/hr ft^2°F.

(a)   0.04
(b)   0.13
(c)   0.35
(d)   0.91
(e)   1.92

Overall coefficient U, thermal conductivity, $\frac{k}{x}$, and film coefficient, h, are

reciprocals of their thermal resistances. Thermal resistances in series are handled analogously to series electrical resistances, hence

$$U \text{ in BTU/(hr ft}^{2\circ}\text{F)} = \frac{1}{\frac{x_1}{k_1} + \frac{x_2}{k_2} + \frac{x_3}{k_3} + \cdots}$$

Overall coefficient U is used in a simplified conduction equation $Q = UA\Delta T$. In this problem:

$$U = \frac{1}{\frac{5/12}{0.80} + \frac{3/12}{0.04} + \frac{1/12}{0.07}} = \frac{1}{0.52 + 6.25 + 1.19} = 0.126 \text{ BTU/(hr ft}^{2\circ}\text{F)} \bullet$$

Answer is (b)

THERMO 88

Given an inner wall at 80°F and an outer wall exposed to ambient wind and surroundings at 40°F. Film coefficient, h, for convective heat transfer at a 15 mph wind is about 7 BTU/(hr ft^{2}°F).

Ignoring any radiation losses calculate an overall coefficienct for the conduction and convection losses.

(a) 0.14
(b) 0.80
(c) 1.25
(d) 7.1
(e) 8.2

Since conduction and convection are based on ΔT, absolute temperatures are not required. For steady state, heat conducted through wall = heat lost by convection:

$$Q = \frac{kA(T_1 - T_2)}{x} = hA(T_2 - T_3) \tag{1}$$

Q can similarly be expressed by an overall coefficient

$$Q = UA(T_1 - T_3) \tag{2}$$

U is calculated in a manner analogous to that used for thermal conductivities in series.

$$U = \frac{1}{\frac{1}{h_1} + \frac{x_1}{k_1} + \cdots} \tag{3}$$

$$U = \frac{1}{\frac{1}{7} + \frac{4/12}{0.30}} = \frac{1}{0.143 + 1.11} = 0.80 \text{ BTU/(hr ft}^{2\circ}\text{F)} \bullet$$

Alternate Solution

If the question had asked for T_2, equation (1) would have given it. Similarly the solution for Q from (1) would yield U from (2):

$$Q = \frac{0.30}{\frac{4}{12}} (80 - T_2) = 7(T_2 - 40)$$

$$Q = 72 - 0.9 T_2 = 7 T_2 - 280 \qquad \therefore T_2 = \frac{352}{7.9} = 44.5°F \quad \text{and} \quad Q = 32 \ \text{BTU}/(\text{hr ft}^2)$$

$$Q = UA(T_1 - T_3) = 32 = U(1)(80 - 40) \qquad U = \frac{32}{40} = 0.80 \ \text{BTU}/(\text{hr ft}^2 °F) \ \bullet$$

Note that this alternate approach becomes inoperative when more than one film coefficient and one conductivity are involved, but lacking information on the interface temperatures.

Answer is (b)

THERMO 89

A metal object at $120°F$ is set on an insulating pad to cool. The temperature initially falls from $120°F$ to $100°F$ in 12 minutes. Surroundings are at $65°F$. Find time required for that object to continue to cool from $98°F$ to $80°F$. Assume negligible conduction and radiation losses in both cases.

rate of heat loss from block = rate of convective heat transfer

$$-\rho CV \frac{d\theta}{dt} = hA\theta \ , \quad \text{where} \ \rho = \text{density}, \ C = \text{heat capacity}, \ V = \text{volume}, \ t = \text{time},$$
$$\theta = \text{temperature difference with surroundings},$$
$$h = \text{convective heat transfer coefficient, and } A = \text{area}.$$

Rearrange and integrate using the first cooling limits:

$$\ln \theta \ \Big]_{\theta_1=55°F}^{\theta_2=35°F} = \left(-\frac{hA}{\rho CV}\right) \ t \ \Big]_{t_1=0\text{min.}}^{t_2=12\text{min.}} \qquad \ln\left(\frac{35}{55}\right) = \left(-\frac{hA}{\rho CV}\right)(12 - 0)$$

Solve for the unspecified constants: $\left(-\frac{hA}{\rho CV}\right) = -.0377 \ \text{min}^{-1}$

Insert the new problem data and re-evaluate the integral. $\theta_3 = 98-65$, $\theta_4 = 80-65$, $t_4 - t_3 =$ unknown.

$$\ln\left(\frac{15}{33}\right) = - .0377 \ (t_4 - t_3) \qquad t_4 - t_3 = 20.9 \ \text{min.} \ \bullet$$

This demonstrates the exponential nature of temperature change with time, where the heat transfer rate by conduction or by convection varies solely with temperature difference, i.e. in the absence of heat generated in, or lost from the object in other ways.

THERMO 90

A shell and tube brine cooler cools 150 gallons of brine per minute from $16°F$ to $12°F$, using ammonia at $5°F$. The effective outside area of the tubes is 310 ft^2. The brine has a specific gravity of 1.2 and a specific heat of 0.70.

(a) Compute the rating of the cooler in tons of refrigeration.

(b) Compute the coefficient of heat transfer, U, in BTU/(hr ft^2°F).

(a) Heat of Cooling = $Q_c = \dot{m}_B\, C_{P_B}(T_1 - T_2)$

where $\dot{m}_B$ = brine mass flow rate

C_{P_B} = brine specific heat = 0.70

T_1 = brine inlet temperature = $16°F$

T_2 = brine outlet temperature = $12°F$

$$\frac{150 \text{ gal}}{\text{min}} \times \frac{\text{cu ft}}{7.48 \text{ gal}} = 20 \text{ ft}^3/\text{min}$$

1 ton of refrigeration = 200 Btu/min = 12,000 Btu/hr

$$\dot{m}_B = \frac{62.4 \text{ lb}}{\text{cu ft}} \times 1.2 \times \frac{20 \text{ cu ft}}{\text{min}} \times \frac{60 \text{ min}}{\text{hr}} = 90,000 \text{ lbm/hr}$$

Q_c = 90,000 x 0.70(16° - 12°) = 252,000 Btu/min

Tons of Refrigeration = $\frac{252,000}{12,000}$ = 21. ●

(b) $U = \dfrac{q}{A_T\, \Delta t_m}$ where q = heat transfer in Btu/hr

A_T = effective area of the tubes in sq ft

Δt_m = log mean temp difference

$$\Delta t_m = \frac{(\Delta t_0)_a - (\Delta t_0)_b}{Ln\left[(\Delta t_o)_a \Big/ (\Delta t_0)_b \right]}$$

$(\Delta t_0)_a = 16° - 5° = 11°F$

$(\Delta t_0)_b = 12° - 5° = 7°F$

$$= \frac{11° - 7°}{Ln(11/7)} = \frac{4°}{0.45} = 8.9°F$$

We are using the logarithmic mean temperature difference which applies to countercurrent heat exchange; ideally a correction should be applied as this is a shell and tube exchanger.

$$U = \frac{q}{A_T\, \Delta t_m} = \frac{252,000 \text{ BTU/hr}}{310 \text{ ft}^2(8.9°F)} = 91.4 \text{ BTU}/(\text{hr ft}^2 °F) ●$$

THERMO 91

Which of the following is <u>not</u> a usual expression of the power/unit area Stefan-Boltzmann constant for black body radiation?

(a) 1.36×10^{-12} cal/(sec cm^2°K^4)

(b) 5.68×10^{-5} ergs/(sec cm^2°K^4)

(c) 5.68×10^{-8} watts/(m^2 °K^4)

(d) 0.171×10^{-8} BTU/(ft^2 hr°R^4)

(e) 5.68×10^{-8} coulombs/(sec m^2°K^4)

All are numerically correct conversion of the constant in terms of power per unit area.

The electrical units of (e) are not recognized as appropriate, although they are numerically equivalent to a correct constant of

$$5.68 \times 10^{-8} \text{ coulombs}/(\text{sec m}^2°\text{K}^4)$$

(e) is <u>not</u> a usual expression ●

8
ELECTRICAL
CIRCUITS

CIRCUITS 1

The equivalent resistance R_{ab} is closest to

(a) 2 ohms
(b) 4 ohms
(c) 6 ohms
(d) 8 ohms
(e) 10 ohms

Solution
R_{eq} for the $8\,\Omega$, $12\,\Omega$, and $16\,\Omega$ resistors in parallel is

$$R_{eq} = \frac{1}{\frac{1}{8} + \frac{1}{12} + \frac{1}{16}} = 3.68\ \Omega$$

and $R_{ab} = 3 + \dfrac{4(2 + 3.68)}{4 + 2 + 3.68} = 5.35\ \Omega\ \bullet$

Answer is c

185

CIRCUITS 2

The voltage V_x is closest to

    (a)   16   volts
    (b)   8   volts
    (c)   3.55 volts
    (d)   6.42 volts
    (e)   4.65 volts

Solution
Using current division

$$i_{8\Omega} = 2 \times \frac{4}{4+14} = \frac{8}{18} \text{ A}$$

and $\quad V_x = 8 i_{8\Omega} = \dfrac{64}{18} = 3.55 \text{ V} \bullet$

Answer is c.

CIRCUITS 3

The current i_x is closest to

    (a)   3.35 amps
    (b)   0.35 amps
    (c)   2.35 amps
    (d)   1.68 amps
    (e)   5.21 amps

Solution
Redrawing the circuit

$$i_T = \frac{10}{.5 + \dfrac{1}{\frac{1}{6} + \frac{1}{3} + \frac{1}{3}}} \quad = 5.87 \text{ A}$$

and

$$i_x = 5.87 \times \frac{2}{2+3} \quad = 2.35 \text{ A} \bullet$$

Answer   is   c.

CIRCUITS 4

The power supplied by the 10 volt source is

    (a)   12 watts
    (b)   0 watts
    (c) -12 watts
    (d)   16 watts
    (e) -16 watts

Solution
Designating the clockwise mesh currents
by i_1 and i_2 and using Kirchhoff's
voltage law we obtain

$$6i_1 - 2i_2 = 10$$
$$-2i_1 + 4i_2 = -20$$

Solving $i_1 = 0$ A and $P_{AV} = 0$. Answer is b●

CIRCUITS 5

The voltage V_y is closest to

    (a)   0 volts
    (b)   3.6 volts
    (c) −1.2 volts
    (d)   7.2 volts
    (e) −7.2 volts

Solution
Denoting the voltage drop (top to bottom)
accros the $2\,\Omega$ resistor as V_1 and using
Kirchhoff's current law, K.C.L., we have

$$\frac{V_1}{2} + \frac{V_1 - 6}{3} = -3$$

Solving, $V_1 = -1.2$

and $\quad V_y = V_1 - 6 = -7.2$ V●

    Answer is e.

CIRCUITS 6

The node voltage V_2 is closest to

(a) 4.32 volts
(b) -4.32 volts
(c) 1.82 volts
(d) -1.82 volts
(e) 0 volts

Solution
Using K.C.L. at nodes 1 and 2 we obtain

$$\left(\tfrac{1}{2}+\tfrac{1}{4}\right)V_1 - \tfrac{1}{4}V_2 = 2$$
$$-\tfrac{1}{4}V_1 + \left(\tfrac{1}{4}+\tfrac{1}{4}+\tfrac{1}{2}\right)V_2 = 1$$

These equations become

$$3V_1 - V_2 = 8$$
$$-V_1 + 4V_2 = 4$$

and $V_2 = \dfrac{20}{11} = 1.82$. Answer is c ●

CIRCUITS 7

The node voltage V_3 is closest to

(a) 2 volts
(b) 4 volts
(c) 6 volts
(d) 8 volts
(e) 10 volts

Solution

We note $V_3 = V_2 \times \dfrac{2}{2+2} = .5\,V_2$

By inspection,

$$V_1 = 6 \quad, \quad V_1 - V_2 = -10 \quad \text{so} \quad V_2 = 16\,V$$

and $V_3 = 8$ Volts● Answer is d.

CIRCUITS 8

The mesh current **i,** is closest to

<table>
<tr><td>(a)</td><td>-3</td><td>amps</td></tr>
<tr><td>(b)</td><td>-5</td><td>amps</td></tr>
<tr><td>(c)</td><td>3</td><td>amps</td></tr>
<tr><td>(d)</td><td>5</td><td>amps</td></tr>
<tr><td>(e)</td><td>2.4</td><td>amps</td></tr>
</table>

Solution

By inspection $i_3 = 2\,A$. Writing K.V.L. for meshes 1 and 2

$$4i_1 - 2i_2 = 12$$
$$-2i_1 + 4i_2 - 2 = 4$$
$$4i_1 - 2i_2 = 12$$
$$-2i_1 + 4i_2 = 6$$

Solving for i_1, $\quad i_1 = 5\,A$ ●

Answer is d.

CIRCUITS 9

The value of R_L to dissipate maximum power is

(a) 0.86 ohms
(b) 7   ohms
(c) 1.43 ohms
(d) 4.26 ohms
(e) 8.76 ohms

Solution
To dissipate maximum power we must make $R_L = R_{TH}$, the Thevenin resistance at the terminals a-b.

$$R_{ab} = \frac{1(2+4)}{1+2+4} = .86 \ \Omega \bullet$$

Answer is a.

CIRCUITS 10

The Thevenin equivalent at terminals **a** - **b** is closest to

(a) $V_T = 5.33$ volts, $R_T = 5$ ohms
(b) $V_T = 5.33$ volts, $R_T = 4.33$ ohms
(c) $V_T = 1.84$ volts, $R_T = 5$ ohms
(d) $V_T = 5.33$ volts, $R_T = 4.33$ ohms
(e) $V_T = 2.67$ volts, $R_T = 4.33$ ohms

Solution
V_T is the open-circuit voltage from a-b.
Therefore, $V_T = 8 \times \dfrac{2}{2+4} = 2.67$ Volts.

R_T is the equivalent resistance with all the sources equal to zero.

$$\therefore \quad R_T = 3 + \dfrac{2 \times 4}{6} = 4.33 \, \Omega \quad .\text{Answer is e.} \bullet$$

CIRCUITS 11

The Thevenin equivalent at the terminals **a - b** is

    (a) **V_T** = 10 volts,   **R_T** = 1.5 ohm
    (b) **V_T** = 12 volts,   **R_T** = 1.0 ohm
    (c) **V_T** = 12 volts,   **R_T** = 1.5 ohm
    (d) **V_T** = 10 volts,   **R_T** = 1.0 ohm
    (e) **V_T** = 0 volts,   **R_T** = 1.0 ohm

Solution
Step 1 - To find V_T
Using nodal analysis $V_1 = 8V$ and K.C.L. at node 2 yields

$$\tfrac{1}{2} V_2 + \tfrac{1}{2}(V_2 - 8) = 2$$

and

$$V_2 = 6 \text{ volts}$$

Using K.V.L

$$V_{ab} = 4 + 6 = 10 \text{ volts}$$

Step 2 - To find R_T
With all of the sources equal to zero
(replace 2A by open)

$$R_T = \frac{2 \times 2}{4} = 1 \, \Omega$$

We note that shorting the 8V source
makes the 4Ω resistor a short. Answer is d ●

CIRCUITS 12

Nortons equivalent at terminals **a** - **b** is

(a) $I_N = 0.2$ amps, $R_N = 4$ ohms
(b) $I_N = 0.2$ amps, $R_N = 5$ ohms
(c) $I_N = -0.5$ amps, $R_N = 4$ ohms
(d) $I_N = -0.5$ amps, $R_N = 5$ ohms
(e) $I_N = -0.5$ amps, $R_N = 2$ ohms

Solution
Step 1 - To find I_N
With the terminals a-b
shorted

$$i_T = \frac{10}{\frac{8}{6} + \frac{48}{14}} = 2.1$$

$$i_1 = 2.1 \times \frac{2}{6} = .70$$

$$i_2 = 2.1 \times \frac{8}{14} = 1.20$$

$$i_N = .7 - 1.2 = -0.5 \; A$$

Step 2 - To find R_N
With the source equal to zero
it can be seen that

$$R_{ab} = \frac{4 \times 6}{10} + \frac{2 \times 8}{10}$$

$$= 4.0 \; \Omega$$

Answer is c. ●

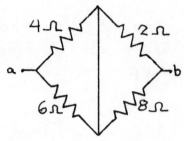

CIRCUITS 13

When solving for i_x using superposition, the contribution due to the 6V source is

(a) -0.24 amps
(b)   0.24 amps
(c)   0 amps
(d)   0.72 amps
(e) -0.72 amps

Solution
Replacing the 2A source by an open circuit
and the 10V source by a short we find

$$i_T = \frac{-6}{7 + \frac{4 \times 2}{6}} = -0.72 \; A$$

and $\quad i_x = -.72 \times \frac{2}{4+2} = -0.24 \; A$ ●

Answer is a.

CIRCUITS 14

Given: $3 \cos(10t) - 4 \sin(10t - 45°) + X(t) = 0.$

X(t) is

(a) 6.48 Cos(10t)
(b)   1 Cos(10t - 135°)
(c) 6.48 Cos(10t + 154°)
(d)   1 Cos(10t + 135°)
(e) 6.48 Cos(10t - 154°)

Solution
Using phasors

$$\bar{X} = -3 \angle 0° + 4 \angle -45° - 90°$$
$$= -3 - 2.83 - j2.83$$
$$= -5.83 - j2.83$$
$$= 6.48 \angle -154°$$

$$\therefore X(t) = 6.48 \cos(10t - 154°). \text{ Answer is e.} \bullet$$

CIRCUITS 15

The root mean square value of X(t) is

(a) 1.6
(b) 3.0
(c) 2.0
(d)   0
(e) 3.8

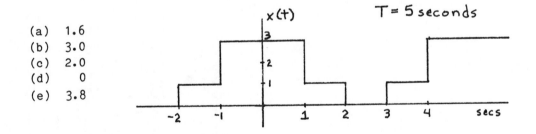

Solution

$$X_{RMS} = \sqrt{\frac{1}{5} \int_{-2}^{3} x^2(t)\, dt}$$

$$= \sqrt{\frac{1}{5}\left[1^2 + 3^2(2) + 1^2 + 0^2\right]}$$

$$= \sqrt{\frac{1}{5}(20)} = 2. \text{ Answer is c.} \bullet$$

CIRCUITS 16

The phasor voltage $\overline{V}_x$ is

(a) 2.17 + j5.78
(b) 6.2 $\angle 110.5°$
(c) 6.2 $\angle 69°$
(d) 6.2 $\angle -69°$
(e) 6.2 $\angle -110.5°$

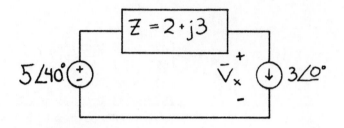

Solution

Using K.V.L. for phasors

$$V_x = -3 [2 + j3] + 5 \angle 40°$$
$$= -6 - j9 + 3.83 + j3.21$$
$$= -2.17 - j5.79$$
$$= 6.2 \angle -110.5° \bullet \quad \text{Answer is} \quad e.$$

CIRCUITS 17

The equivalent impedance Z_{ab} is

(a)   0.62 - j1.55 ohms
(b)   0.62 + j1.55 ohms
(c) -0.62 + j0.45 ohms
(d)   0.62 - j0.45 ohms
(e) -0.62 + j0.45 ohms

Solution

$$Z_{ab} = -j + \cfrac{1}{\frac{1}{2} + \frac{1}{2+j} + \frac{1}{-j}}$$

$$= -j + \frac{1}{.5 + .4 - j0.2 + j} \quad = -j + \frac{1}{0.9 + j0.8}$$

$$= -j + \frac{1}{1.20 \underline{/41.6°}} \quad = -j + 0.83 \underline{/-41.6°}$$

$$= -j + 0.62 - j0.55 = 0.62 - j1.55 \quad \Omega \bullet$$

Answer is a .

CIRCUITS 18

The impedance or admittance at **a** – **b** is real when the reactance X_c is

(a)   2 – j4   ohms
(b)   – j5   ohms
(c)   0.2   ohms
(d)   5   ohms
(e)   0   ohms

Solution

We require that $j\frac{1}{X_c} + \frac{1}{2+j4}$ is real

or $j\frac{1}{X_c} + 0.1 - j0.2 = 0.1 + j0$

$\therefore \quad \frac{1}{X_c} = 0.2 \quad$ or $\quad X_c = 5 \quad \Omega \bullet$

Answer is d .

CIRCUITS 19

V(t) is

    (a)   1.13 Cos(2t + 55.3°)
    (b)   2.24 Cos(2t - 78°)
    (c)   1.13 Cos(2t - 55.3°)
    (d)   2.24 Cos(2t + 78°)
    (e)   2.6 Cos(2t)

Solution
Considering the phasor-impedance circuit

$$Z_1 = \frac{1}{1+j} = 0.5 - j0.5$$

By phasor voltage
division

$$\bar{V} = 4\,\frac{0.5 - j0.5}{2.5 - j0.5}$$

$$\therefore \quad \bar{V} = \frac{4\angle0°\ 0.71\angle-45°}{2.55\angle11.3°} = 1.13\angle-55.3°$$

and $V(t) = 1.13\ Cos\ (2t - 55.3°)$. Answer is c. ●

CIRCUITS 20

i(t) is

    (a)   -0.5 Cos(2t)   amps
    (b)   0.5 Cos(2t)   amps
    (c)   Sin(2t)   amps
    (d)   0.5 Cos(2t) + 0.5 Sin(2t)   amps
    (e)   Cos(2t)   amps

Solution
Considering the phasor
impedance circuit

$$\bar{I}_T = \frac{-j}{j2 + \dfrac{1}{j0.5 - j0.25}} = \frac{-j}{j2 - j4}$$

$$= .5$$

By current division $I = 0.5 \dfrac{-j2}{-j2 + j4} = -0.5$

$\therefore \; i(t) = -0.5 \cos 2t \; A.$ Answer is   a .●

CIRCUITS 21

I_x is

(a)   2.7 − j6.2   amps
(b)   2.7 + j6.2   amps
(c)   2 + j3   amps
(d)   2 − j3   amps
(e)   5.4 $\underline{/-68°}$ amps

Solution
Since there is a short from node 2 to the
reference $\bar{V}_2 = 0$ and $\bar{V}_1 = 3$.
By K.C.L.

$$\bar{I}_x = (2+j) - j3 + \frac{3}{j}$$

$$= 2 - j5 = 5.4 \underline{/-68°} . \text{Answer is e .●}$$

CIRCUITS 22

The voltage **V(t)** is

(a)   5.33 Cos(2t)   volts
(b)   0   volts
(c)   -2 + 5.33 Cos(2t)   volts
(d)   2 + 5.33 Cos(2t)   volts
(e)   2 + 5.12 Cos(2t - 12°) volts

━━━━━━━━

Solution
Using superposition for the two sources

$$i_T = \frac{6}{2+1} = 2A$$

$$v(t) = -2 \text{ volts}$$

$$\bar{V} = 8\angle 0° \frac{2}{3} \text{ (Be sure)}$$

$$= 5.33 \angle 0°$$

$$\therefore \ v(t) = -2 + 5.33 \ \text{Cos} \ 2t \ V. \text{Answer is c.} \bullet$$

CIRCUITS 23

The power supplied by the current source is closest to

(a)   0 watts
(b)   5 watts
(c)   10 watts
(d) -15 watts
(e) -20 watts

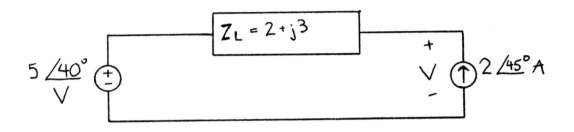

Solution
$\bar{V} = (2+j3)2 \angle 45° + 5 \angle 40° = (3.6 \angle 56.3°)(2 \angle 45°) + 5 \angle 40°$
$= 7.2 \angle 101.3° + 3.83 + j3.21 = -1.37 + j7.06 + 3.83 - j3.21$
$= 2.46 + j3.85 = 4.57 \angle 57°$

$P_{av} = \frac{1}{2} V_m I_m \cos(\phi_v - \phi_i) = (4.57)(2) \cos(57° - 45°)$
$= 8.94$ watts ● Answer is c.

CIRCUITS 24

The power dissipated by Z_L in the impedance of the previous problem is

    (a)   2 watts
    (b)   4 watts
    (c)   6 watts
    (d)   8 watts
    (e) 10 watts

Solution

$I_{rms} = 2$

$\therefore \quad P_{av} = I_{rms}^2 \, \mathrm{Re}[2 + j3]$
$= 4 \times 2$
$= 8$ watts ●

Answer is d.

CIRCUITS 25

If 30 kVA is supplied to a load of 30 amps at a power factor
of 0.7 lagging, what is the reactive part of the load?

(a)   -22.7   ohms
(b)   22.7   ohms
(c)   6.4   ohms
(d)   -6.4   ohms
(e)   0   ohms

Solution

The power diagram is shown.

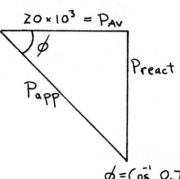

$$P_{app} = \frac{20}{.7} = 28.5 \quad \text{and}$$

$$P_{react} = 28.7 \sin 45.5°$$
$$= 20.5 \text{ kVar}$$

$$\therefore \quad 20.5 \times 10^3 = (30)^2 X_c \quad \text{and}$$

$$X_c = \frac{20.5 \times 10^3}{900} = 22.7 \quad \text{so} \quad Z_L = R - j22.7 \bullet$$

$\phi = \cos^{-1} 0.7$
$\phi = 45.5°$

Answer is a.

CIRCUITS 26

The real power dissipated in the load is closest to

(a)   50   watts
(b)   100   watts
(c)   150   watts
(d)   200   watts
(e)   250   watts

Solution

$$Z_L = \frac{20(-j10)}{20-j10} = \frac{200\angle -90°}{22.3\angle 26.6°} = 9.0\angle -63.4° = 4.02 - j8.04$$

$$I_{RMS} = \frac{100}{|5+4.02-j8.04|} = \frac{100}{12.1} = 8.26 \text{ amps}$$

$$P_{LOAD} = I_{RMS}^2 \text{ Re}[Z_L] = (8.26)^2 4.02$$
$$= 274 \text{ watts.} \bullet \qquad Answer \text{ is } e.$$

CIRCUITS 27

Given a 440 volt line supplies 20 kWatts at a power factor of
0.8 lagging, the line current is closest to

(a) 20 amps
(b) 40 amps
(c) 60 amps
(d) 80 amps
(e) 100 amps

Solution
The apparent power is
$$P_{app} = \frac{20 \times 10^3}{.8} = 25 \times 10^3 \text{ watts}$$

$$P_{app} = V_{RMS} I_{RMS}$$

$$\therefore \quad I_{RMS} = \frac{25\times10^3}{440} = 56.8 \text{ A} \bullet$$

Answer is c.

$\phi = Cos^{-1} 0.8$
$= 36.8°$

CIRCUITS 28

The value of C so the load has unity power factor at 200
rads/sec is closest to

(a) 100 microfarads
(b) 200 microfarads
(c) 300 microfarads
(d) 400 microfarads
(e) 500 microfarads

Solution

$P_{react. load} = 20 \times 10^3 \tan 36.8° = 14.96 \ kVar$
Therefore we require the reactive power for $C = 14.96 \ kVar$

$$\frac{V_{RMS}}{X_c^2} X_c = 14.96 \times 10^3$$

$$\therefore X_c = \frac{400^2}{14.96 \times 10^3} = 10.7 \ \Omega \ ; \quad X_c = \frac{1}{\omega C}$$

$$\frac{1}{200 C} = 10.7 \ \Omega \quad or \quad C = \frac{10^6}{200 \times 10.7} \ \mu F$$

This yields $C = 467 \mu F$ ●

Answer is e.

CIRCUITS 29

The forced voltage V_c is

(a) 1.14 volts
(b) −1.14 volts
(c) 2.28 volts
(d) −2.28 volts
(e) 0 volts

Solution
For D.C. forced response we replace a capacitor by an open circuit.

$$\therefore \quad V_c = \frac{2 \times 4 \times 2}{4+3}$$

$$= 2.28 \text{ volts. Answer is c.} \bullet$$

CIRCUITS 30

A long time after the switch moves from position 1 to 2, i_1 is

(a) 3.33 amps
(b) 2.0 amps
(c) 2.5 amps
(d) 4.3 amps
(e) 0   amps

Solution
For the D.C. forced response we have the circuit shown with the inductor short circuited.

$$i_1 = \frac{10}{3+1} = 2.5 \text{ amps} \bullet$$

Answer is c.

CIRCUITS 31

The complete response for **i(t)** at **t>0** is

(a) $2.5 + Ae^{-0.75t}$

(b) $2.5 - 2.5e^{-0.75t}$

(c) $2.5 + 2.5e^{-1.33t}$

(d) 0

(e) $2.5 - 2.5e^{-1.33t}$

assume $i(0) = 0$

Solution

$$3\frac{di}{dt} + 4i = 10$$

$$i_{tr}(t) = A e^{-1.33t}$$

$$i_{fo}(t) = 2.5$$

$$i_{com}(t) = 2.5 + A e^{-1.33t}$$

Applying $i(0) = 0$

$$0 = 2.5 + A$$
$$A = -2.5$$

$$i_{com}(t) = 2.5 - 2.5 e^{-1.33t} \qquad \bullet \text{ Answer is e.}$$

CIRCUITS 32

The form of the transient part of **V(t)** for **t>0** is

(a) $(A_1 + A_2)e^{-0.5t}$

(b) $A_1 \cos 0.5t + a_2 \sin 0.5t$

(c) $A_1 e^{-t}\cos 0.5t + a_2 e^{-t}\sin 0.5t$

(d) $A_1 e^{-1.71t} + A_2 e^{-0.29t}$

(e) 0

At t=0 switch moves from 1 to 2

(Note: This circuit is also used for the next three problems.)

Solution

For t>0 $\frac{di}{dt} + 2i(t) + \frac{1}{2}\int_0^t i(\alpha)d\alpha + v_c(0^+) = 5\cos t$

$i''(t) + 2i'(t) + .5i(t) = -5\sin t$

The characteristic equation is

$m^2 + 2m + 0.5 = 0$

$(m+1)^2 + 0.5 - 1 = 0$

$\therefore \quad (m+1)^2 = 0.5$

and $m = -1 \pm \sqrt{0.5} = -1 \pm 0.71 = -1.71, -0.29$

i(t) or v(t) has a transient with the form

$v(t) = A_1 e^{-1.71t} + A_2 e^{-0.29t}$ ●

Answer is d.

CIRCUITS 33

The forced part of **v(t)** for **t>0** in the circuit in **CIRCUITS 32** is

    (a) 4 Sin t + 2 Cos t

    (b) 4.85 Cos(t + 14°)

    (c) 2.82 Cos(t + 36°)

    (d) 2.82 Cos(t - 36°)

    (e) 4.85 Cos(t - 14°)

Solution
The phasor impedance
circuit for $\overline{V}$ is

$$\overline{V} = \frac{5}{2 + j0.5} \times 2 = \frac{10}{2.06\angle 14°}$$

$$= 4.85 \angle{-14°} \quad \therefore \quad v(t) = 4.85 \cos(t-14°). \text{ Answer is e.} \bullet$$

CIRCUITS 34

Refer again to the circuit presented in **CIRCUITS 32.**
Assuming steady-state conditions when switching occurs, **V(0⁺)**
is

        (a)   8 volts
        (b)   5 volts
        (c)   3 volts
        (d)   0 volts
        (e) -5 volts

Solution
At $t = 0^-$ before switching $i_L(0^-) = 0$ and $v_c(0^-) = 8V$
The circuit at $t = 0^+$ just after switching is

Since $\quad i(0^+) = 0$

then $\quad v(0^+) = 0 \bullet$

      Answer is d.

CIRCUITS 35

Refer again to the circuit in **CIRCUITS 31.**

$\dfrac{dV(0^+)}{dt}$ in the circuit is

      (a) -6 volts/sec
      (b)   3 volts/sec
      (c)   0 volts/sec
      (d)   6 volts/sec
      (e) -3 volts/sec

Solution
Considering the circuit at
$t = 0^+$ by K.V.L. we have
$v_L(0^+) = 5 - 0 - 8 = -3$ volts

$\therefore \ 1\dfrac{di(0^+)}{dt} = -3$

and $\dfrac{dv(0^+)}{dt} = 2\dfrac{di(0^+)}{dt} = -6$ ● Answer is a.

CIRCUITS 36

i(t) is

    (a) $e^{-1.67t}$

    (b) $e^{-1.67t}$

    (c) $-e^{-1.67t}$

    (d) $-e^{-0.6t}$

    (e) $2 - e^{-1.67t}$

$u(t) = 0, \ t < 0$
$\qquad = 1, \ t > 0$

Solution
The time-constant for the transients is found
with the source short-circuited. $R_{eq} = 5\,\Omega$ and
$$i(t) = A e^{-\frac{1}{5(1/3)}t} = A e^{-0.6t}$$

The forced current $i_{fo}(t) = 0$

$\therefore \qquad i(t) = A e^{-0.6t}$

To find $i(0^+)$ we
consider
$2i_1(0^+) + 2i_1(0^+) - 2i(0^+) = 10$
or
$4i_1(0^+) - 2i(0^+) = 10$ (1) and for mesh 2 we have

$-2i_1(0^+) + 6i(0^+) = 0$ (2). Solving (1) and (2)

$i(0^+) = 1$ and $\frac{1}{3}\frac{dv(0^+)}{dt} = 1$ or $\frac{dv(0^+)}{dt} = 3$.

Answer is b. ●

CIRCUITS 37

Given $i(0^+) = 0$,

$\frac{di(0^+)}{dt}$ is

(a) 0 amps/sec
(b) -2.5 amps/sec
(c) 2.5 amps/sec
(d) 1.25 amps/sec
(e) -1.25 amps/sec

Solution
The circuit at $t = 0^+$ is

Kirchhoff's current law yields at
node 1
$$\tfrac{1}{2} v_1(0^+) + \tfrac{1}{4}(v_1(0^+) - v_2(0^+)) = 5 \quad \text{or} \quad 3v_1(0^+) - v_2(0^+) = 20 \quad \textbf{(1)}$$
node 2
$$\tfrac{1}{2} v_2(0^+) + \tfrac{1}{4}(v_2(0^+)) - v_1(0^+) = 0 \quad \text{or} \quad -v_1(0^+) + 3v_2(0^+) = 0 \quad \textbf{(2)}$$
Solving for $v_2(0^+)$ yields $v_2(0^+) = 2.5$.

$$\therefore \quad 2\frac{di(0^+)}{dt} = 2.5 \quad \text{and} \quad \frac{di(0^+)}{dt} = 1.25 \text{ amps/sec} \; \bullet$$

Answer is d.

CIRCUITS 38

The transient part of $Y(t)$ in the solution of
$Y''(t) + 3Y'(t) + 4Y(t) = 0$ is

     (a)   underdamped
     (b)   overdamped
     (c)   critically damped
     (d)   oscillatory damped
     (e)   constant

Solution
The characteristic equation is
$$m^2 + 3m + 4 = 0$$
$$(m + 1.5)^2 + 4 - 1.5^2 = 0$$
$$\therefore \quad (m + 1.5)^2 = -1.75$$
$$m = -1.5 \pm j1.32$$

and
$$y(t) = A_1 e^{-1.5t} \cos 1.32t + A_2 e^{-1.5t} \sin 1.32t$$

This response is underdamped. $\bullet$
    Answer is a.

CIRCUITS 39

The forced part of **i(t)** is

(a)   0.5 Cos t + Sin t

(b)   0.16 Cos(2t - 72°)

(c)   0.25 + 0.16 Cos(2t - 72°)

(d)   0

(e)   0.25 + 1.6 Cos(2t - 72°)

Solution

$$Cos^2 t = 0.5 + 0.5 \, Cos \, 2t$$

Using superposition we have

$$i_{f_0}(t) = 0.25$$

$$\overline{I} = \frac{0.5 \angle 0°}{2 + j6} = \frac{0.5 \angle 0°}{6.32 \angle 72°} = .16 \angle -72°$$

∴ $i_{f_0}(t) = 0.25 + 0.16 \, Cos \, (2t - 72°)$. Answer is c. ●

CIRCUITS 40

Design a parallel RLC circuit with Q = 10, a resonant frequency $f_0 = 10^5$ Hz and a response of 50 volts to 30 m amps at resonance.

(a)   Not enough information.

(b)   R = 1.67 k-ohms,   L = 0.84 mH,   C = 0.95 x 10^{-8} farads

(c)   R = 0.004 ohms,   L = 20 mH,   C = 2.64 microfarads

(d)   R = 1.67 k-ohms,   L = 20 nH,   C = 0.95 x 10^{-8} farads

(e)   R = 1.67 k-ohms,   L = 0.84 mH,   C = 2.64 microfarads

Solution
For parallel resonance $V_{RMS} = \left| \dfrac{I_{RMS}}{G + j(\omega C - \frac{1}{\omega L})} \right|$,

$$\omega_0 = \frac{1}{\sqrt{LC}} \quad \text{and} \quad Q = \omega_0 RC$$

We are given $50 = \dfrac{30 \times 10^{-3}}{G}$ $\therefore R = \dfrac{50}{30 \times 10^{-3}} = 1670\,\Omega$,

$C = \dfrac{10}{2\pi \times 1670} = 0.95 \times 10^{-8}$ farads, $L = \dfrac{1}{4\pi^2 \times 10^{10} \times 0.95 \times 10^{-8}}$ H

$\therefore L = 0.84$ m henries

Answer is b. ●

CIRCUITS 41

There is a point charge of 0.3 mC at the origin. What is the
magnitude of the electric field intensity at a point located 2
meters in the X-direction, 3 meters in the Y-direction, and 4
meters in the Z-direction away from the origin?

    (a) 500 kV/m
    (b)   5 kV/m
    (c)   93 kV/m
    (d) 9.3 MV/m
    (e) 1.2 MV/m

Solution,
$$\vec{E} = \frac{Q}{4\pi \epsilon_0 R^2} \vec{a}_R$$

The electric field intensity magnitude is
$$E = \frac{Q}{4\pi \epsilon_0 R^2} \quad \text{where} \quad R = \sqrt{2^2 + 3^2 + 4^2}$$

so
$$E = \frac{0.3 \times 10^{-3} \ C}{(4\pi)(8.85 \times 10^{-12} \ F/m)(29 m^2)} = 93 k \frac{V}{m} \ \bullet$$

Answer is c.

CIRCUITS 42

The infinite sheet of charge, with positive charge density, σ, has an electrix field

$$\vec{E} = \frac{6}{2\epsilon_o} \vec{ax} \quad \text{for } x > b.$$

If a second sheet of charge with charge density $-\sigma$ is then placed at a, what is the electric field for a < x < b?

(a) $\frac{6}{\epsilon_o} \vec{ax}$

(b) 0

(c) $-\frac{6}{\epsilon_o} \vec{ax}$

(d) $\frac{6}{2\epsilon_o} \vec{ax}$

(e) $-\frac{6}{2\epsilon_o} \vec{ax}$

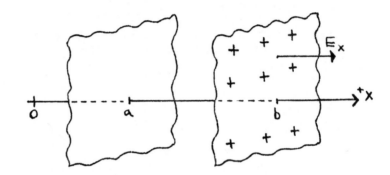

Solution

The E field for the positive sheet of charge alone is $\quad \vec{E}_+ = \frac{-\sigma}{2\epsilon_o} \vec{ax} \quad$ for $\quad x < b$.

And the E field for the negative sheet of charge alone is

$$\vec{E}_- = \frac{-6}{2\epsilon_o} \vec{ax} \quad \text{for} \quad x > a$$

The field between the two sheets is then

$$\vec{E} = \left(\frac{-6}{2\epsilon_o}\right) \vec{ax} + \left(\frac{-6}{2\epsilon_o}\right) \vec{ax} = \frac{-6}{\epsilon_o} \vec{ax} \quad \bullet$$

Answer is c.

CIRCUITS 43

The two parallel wires carry constant positive currents in the same direction as shown. These two wires

(a)   attract each other and the magnetic field at point C due to wire **a** is in the positive Z-direction.

(b)   repel each other and the magnetic force on wire **c** due to wire **a** points in the positive Y-direction.

(c)   attract each other and the magnetic force on wire **a** due to wire **c** points in the negative Y-direction.

(d)   repel each other and the magnetic field on wire **a** due to wire **c** is cancelled by the magnetic field due to the current in wire **a**.

(e)   attract each other and the magnetic field at point A due to wire **c** is in the positive Z-direction.

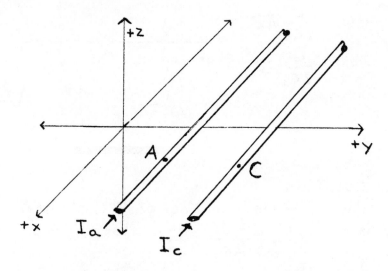

Solution
The equation $F = i\,\ell \times B$ gives the magnetic force directions. The magnetic field directions are found by using the "right-hand rule" for current in a wire.
From the diagram it can be seen that the answer is e. ●

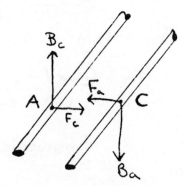

CIRCUITS 44

The electric field between the two capacitor plates is

$$E = \frac{+q}{2\pi \epsilon_o r l}$$

Assuming that $l \gg b$, find the capacitance in terms of a, b, and l.

(a) $\dfrac{2\pi \epsilon_o l}{a-b}$

(b) $\dfrac{2\pi \epsilon_o l}{b^2-a^2}$

(c) $\dfrac{2\pi \epsilon_o l}{b-a}$

(d) $\dfrac{2\pi \epsilon_o l}{\ln \frac{b}{a}}$

(e) $\dfrac{2\pi \epsilon_o l}{e^{(b-a)}}$

Solution

The potential difference between the plates is

$$V = -\int_a^b \vec{E} \cdot \vec{dl} \quad \text{and} \quad \vec{E} = \frac{q}{2\pi \epsilon_o r l} -\vec{a_r} \quad , \quad \vec{dl} = dr\, \vec{a_r}$$

$$V = -\int_a^b \frac{-q}{2\pi \epsilon_o r l}\, \vec{a_r} \cdot dr\, \vec{a_r} = \int_a^b \frac{q}{2\pi \epsilon_o l r}\, dr$$

$$V = \frac{q}{2\pi \epsilon_o l} \ln \frac{b}{a}$$

The capacitance, $C = \dfrac{q}{V}$

$$\therefore C = \frac{2\pi \epsilon_o l}{\ln \frac{b}{a}} \bullet \qquad . \quad \text{Answer is d.}$$

9
ELECTRONICS AND ELECTRICAL MACHINERY

ELECTRONICS

ELECTRONICS 1

Assume that the op-amp is ideal. Find V_o.

(a) 4 v
(b) 6 v
(c) 8 v
(d) $+V_s$
(e) $-V_s$

Solution

Since the op-amp is ideal the virtual short principle can be applied. Thus,

$$V_+ = V_- = 4 \text{ V} \quad \text{and} \quad i_{2k} = \frac{4V}{2k\Omega} = 2\,mA$$

$$i_{1K} = i_{2K} \quad \text{so} \quad i_{1K} = 2\,mA$$

$$V_o = 2\,mA\,[1k + 1k] + 4 = 8 \text{ V} \bullet$$

Answer is c.

ELECTRONICS 2

Assume that the diode is ideal. Find the average current **i**.

(a) -2.12
(b) -0.95
(c)   0
(d) +0.95
(e) +2.12

Solution

Current flows through the diode during half of each voltage cycle. The peak value of the current is

$$\frac{-12 \text{ V}}{4 \text{ k}\Omega} = -3 \text{ mA}$$

The average current is

$$i_{av} = \frac{3}{2\pi} \int_{\pi/6}^{\pi/2} 3 \cos 3t \, dt \quad \text{mA}$$

$$= \frac{3}{2\pi} \left(\sin \frac{3\pi}{2} - \sin \frac{\pi}{2} \right) = -\frac{3}{\pi} = -.95 \text{ mA} \bullet$$
$$\text{Answer is } b.$$

ELECTRONICS 3

The saturation current for the diode is 1 x 10⁻¹⁰ A. The thermal voltage is 25 mV. The i - v relationship can be approximated by

$$i = I_s e^{v/2V_T}$$

Find **v**.

(a) 0.70 V
(b) 0.75 V
(c) 0.80 V
(d) 0.85 V
(e) 0.90 V

Solution

The current is $i = \dfrac{10-v}{4k}$. From the equation

$i = I_s e^{v/2V_T}$ we can solve for v.

$v = 2V_T \ln \dfrac{i}{I_s}$. Using an iterative method to

solve these equations and an initial guess of $v = .7$
we have $i = \dfrac{10-.7}{4k} = 2.33 \ mA$

$v = 2(25mV) \ln \dfrac{2.33 \ mA}{1 \times 10^{-10}} = .85$, $i = \dfrac{10-.85}{4k} = 2.29mA$

$v = 50mV \ln \dfrac{2.29 \ mA}{1 \times 10^{-10}} = .85 \ V$ ● Answer is d.

ELECTRONICS 4

The dc power supply voltages are +12V and −12V. The output
signal, **V₀** is most nearly

    (a)   A sine wave with a peak value of 12V.
    (b)   An unsymmetrical square wave that stays mostly at +12v.
    (c)   A symmetrical square wave that varies between +12 and −12V.
    (d)   An unsymmetrical square wave that stays mostly at −12v.
    (e)   A sine wave with a peak-to-peak voltage of 11v.

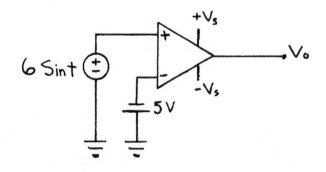

Solution
Since there is no negative feedback the output
will be either +12 V or -12 V. When the
positive input terminal is greater than -5 V, the output is +12 V. When it is less than -5 V, then V_0 is -12.

Answer is b. ●

ELECTRONICS 5

The op-amp has a gain of 100. No current flows through the
op-amp V^+ and V^- terminals. Find V_0/V_{in}.

(a)   -1.94
(b)   -2.00
(c)   -2.06
(d)   +2.00
(e)   +2.06

Solution
$V_0 = 100(V_+ - V_-)$ but $V_+ = 0$ so $V_0 = -100 V_-$

$i_{2k} = \dfrac{V_0 - V_{in}}{6K}$ so $V_- = V_{in} + 2k\left(\dfrac{V_0 - V_{in}}{6k}\right)$

∴ $V_0 = -100 V_{in} - \dfrac{100}{3}(V_0 - V_{in})$

$\dfrac{V_0}{V_{in}} = \dfrac{-66.67}{34.33} = -1.94$ ●

Answer is a.

ELECTRONICS 6

The transistor is an n-channel JFET. The pinch-off voltage is
-2.5V. The saturated drain-to-source current, I_{DSS}, is 12 mA.
Find V_D.

(a) 2 V
(b) 15 V
(c) 5 V
(d) 0 V
(e) 3 V

$+15\ V$

$1k\ \Omega$

V_D

Solution
Assume the transistor is in the saturation
region. The current is

$$I_D = I_{DSS}\left(1 - \frac{V_{GS}}{V_P}\right)^2$$

$$= 12\ mA\left(1 - \frac{0}{-2.5}\right)^2 = 12\ mA$$

$$V_D = 15 - (1k)(12\ mA) = 3\ V \ \bullet$$

Check for saturation region: $V_{GD} = 0 - 3\ V$

$$V_{GD} < V_P$$

Answer is e. $-3 < -2.5$

ELECTRONICS 7

The transistor is an n-channel JFET. Assume r_o approaches
infinity. Use a first-order small-signal model to find R.
The transconductance, g_m, is 3 mA/V.

(a) 444 ohms
(b) 200 ohms
(c) 333 ohms
(d) 500 ohms
(e) 190 ohms

V_i

$500\ \Omega$ R $4k\ \Omega$

Solution

Replacing the JFET with a small-signal model gives the circuit:

Using Thévenin's theorem between node 1 and ground gives open circuit voltage —

$g_m V_{gs} = 0$ so $V_{gs} = 0$

∴ $V_{TH} = V_i$

short circuit current —

$i_{sc} = g_m V_{gs} = g_m V_i$

The equivalent circuit is

$$R_{TH} = \frac{V_i}{g_m V_i} = \frac{1\,V}{3mA} = 333\,\Omega$$

∴ $R = 333 // 500 = 200\,\Omega$ ● Answer is b.

ELECTRONICS 8

The transistor is an npn BJT. The common-emitter current gain, β is 40. Assume the forward-bias voltage V_{BE} equals 0.7V. Find V_C.

(a) 15   V
(b) 13.4 V
(c) 0   V
(d) 14.5 V
(e) 13.7 V

Solution
Assume the transistor is in the active mode.

$V_E = 4 - v_{BE} = 4 - .7 = 3.3$ V, $\quad I_E = \dfrac{3.3 \text{ V}}{5 K \Omega} = .66 \text{ mA}$

$I_c = \alpha I_E = \dfrac{\beta}{\beta + 1} I_E = \dfrac{40}{41} (0.66 \text{ mA}) = 0.64 \text{ mA}$

$V_c = 15 - 2 K \Omega (0.64 \text{ mA}) = 13.72$ V $\bullet$

The collector voltage is higher than the base voltage by 9.72 volts, so the active mode assumption was correct.
Answer is e.

ELECTRONICS 9

The transistor is an npn BJT, with $V_{BE} = 0.7$V and $\beta = 100$.
Find I_B.

    (a)   0.133 Milliampere
    (b)   0 Amp
    (c) 12.7 Microampere
    (d) 25.4 Microampere
    (e)   2.54 Milliampere

Solution

Using Thévenin's theorem the base circuit becomes:

$$V_B = 20V \left(\frac{50K}{50K + 100K} \right)$$

V_B ⊥— ⌇⌇ R_B

$$V_B = 6.67 \text{ V} \quad \text{and} \quad R_B = 100K // 50K = 33.3 K\Omega$$

I_B can be found by writing a loop equation.

$$6.67 \text{ V} = I_B R_B + V_{BE} + (2K) I_E$$
$$6.67 = I_B 33.3K + .7 + 2K(\beta + 1) I_B$$
$$I_B = \frac{6.67 - .7}{33.3K + 2K(101)} = 25.4 \mu A. \text{ Answer is d.} \bullet$$

ELECTRONICS 10

The transistor is an enhancement MOSFET with $\beta = 2 \text{ mA/V}^2$ and $V_T = 2V$. Find V_D.

(a) 8 V
(b) -7 V
(c) 6 V
(d) 0 V
(e) 5 V

Solution

Using the voltage divider rule at the gate gives:

$$V_G = 20 \, V \left(\frac{100k}{100k + 150k} \right) = 8 \, V$$

Assume the device is in the saturation mode.

$$I_D = \frac{1}{2} \beta (V_{GS} - V_T)^2 \quad ; \quad V_{GS} = V_G - (1k)(I_D)$$

$$I_D = \frac{1}{2} (2)(V_G - I_D - 2)^2 \quad mA$$

$$I_D = (6 - I_D)^2 = I_D^2 - 12 I_D + 36 \quad mA$$

$$0 = I_D^2 - 13 I_D + 36$$

Solving this equation gives $I_D = 9 mA$ or $4 mA$

For $I_D = 4 mA$, $V_{GS} = 8 - 4 = 4 \, V$

Checking for the saturation mode we have

$$V_{GS} > V_T$$
$$4 \, V > 2 \, V$$

and $V_D = 20 - 3k(4mA) = 8 \, V$, $V_{GD} = 8 - 8 = 0 \, V$

$$V_{GD} < V_T$$
$$0 < 2 V$$

∴ The transistor is in the saturation mode. Answer is a. ●

ELECTRONICS 11

The transistor is an n-channel JFET. I_{DSS} is 16 mA and V_p is
-4V. Use a first-order small signal model and let r_o approach
infinity in Part 3.

Part (1). Find V_{GS}.

(a) -6.6 V
(b) -0.7 V
(c) -1.9 V
(d) 0 V
(e) -2.7 V

Solution
Using the voltage divider rule at the gate gives
$$V_G = 20 \, V \left(\frac{250k}{250k + 500k} \right) = 6.67 \, V$$

Then $V_{GS} = 6.67 - 2k(I_D)$
Using the equation $I_D = I_{DSS} \left(1 - \frac{V_{GS}}{V_P} \right)^2$ gives

$$I_D = 16 \left(1 - \frac{6.67 - 2I_D}{-4} \right)^2$$

$$I_D = 113.85 - 42.68 \, I_D + 4 \, I_D^2$$
$$0 = I_D^2 - 10.92 \, I_D + 28.46$$

Solving this gives $I_D = 4.30$ mA or 6.62 mA
For the saturation region $V_{GS} > V_P$ so $I_D = 4.3$ mA
$\therefore V_{GS} = 6.67 - 2k(4.3 \, mA) = -1.93 \, V$. Answer is c. ●

Part (2). Find g_m.

    (a)   4.1 mA/V
    (b) −4.1 mA/V
    (c)   5.2 mA/V
    (d) −5.2 mA/V
    (e)   3.3 mA/V

Solution

$$gm = \frac{dI_D}{dV_{GS}} = \frac{d}{dV_{GS}}\left(I_{DSS}\left(1 - \frac{V_{GS}}{V_P}\right)^2\right)$$

$$= \frac{2I_{DSS}}{-V_P}\left(1 - \frac{V_{GS}}{V_P}\right)$$

$$= \frac{2(16)}{4}\left(1 - \frac{-1.93}{-4}\right) = 4.14 \frac{mA}{V} \bullet$$

Answer is a.

Part (3). Find the gain V_o/V_i at mid-band frequencies.

(a) -3.8
(b) +3.8
(c) -6.15
(d) +6.15
(e) -4.4

Solution
The small-signal model in the circuit is:

$$V_{gs} = V_i\left(\frac{250/\!/500}{250/\!/500 + 100}\right) = .625 \quad ; \quad \frac{V_{gs}}{V_i} = .625$$

$$V_o = -gm V_{gs}\left(2k/\!/6k\right)$$

$$\frac{V_o}{V_{gs}} = -4.14\,mA\left(1500\,\Omega\right) = -6.15$$

$$\frac{V_o}{V_i} = \frac{V_o}{V_{gs}} \times \frac{V_{gs}}{V_i} = -6.15 \times .625 = -3.84 \bullet$$

Answer is a.

ELECTRONICS 12

The two transistors are n-channel enhancement MOSFET's. $V_{T1} = 1.83V$. $V_{T2} = 3.00V$ and $\beta_1 = 0.5 \text{ mA/V}^2$. $I_{D1} = 3.375 \text{ mA}$.

Part 1. For both transistors to be in the saturation mode, what must β_2 be?

   (a) 0.33 mA/V^2
   (b) 0.44 mA/V^2
   (c) 0.55 mA/V^2
   (d) 0.66 mA/V^2
   (e) 0.77 mA/V^2

Part 2. What is V_o?

   (a) 6.5 V
   (b) 7.0 V
   (c) 7.5 V
   (d) 8.0 V
   (e) 6.0 V

Solution

The current through both transistors must be equal and satisfy the equation $I_D = \frac{1}{2} \beta (V_{GS} - V_T)^2$.

$$I_{D1} = I_{D2} = \frac{1}{2} \times \frac{1}{2} (12 - V_o - 1.83)^2 = \frac{1}{2} \beta_2 (V_o - 3)^2$$

Solving these two equations for V_o and β_2 we get
$V_o = 6.5 \text{ V} \bullet$ from $\quad 3.375 = \frac{1}{4} (10.17 - V_o)^2$
and
$\beta_2 = 0.551 \frac{mA}{V^2} \bullet$ from $\quad 3.375 = \frac{\beta_2}{2} (6.5 - 3)^2$

Part 1, Answer is c.
Part 2, Answer is a.

MACHINERY

MACHINERY 1

For the cast iron core below, l_1 = 1.3 m, l_2 = 0.4 m, the field intensity in part 1, H_1 = 700 At/m, and the field intensity in part 2, H_2 = 600 At/m.

Find **i**.

(a) 1.7 A
(b) 2.3 A
(c) 2.6 A
(d) 1.8 A
(e) 2.8 A

Solution

Using an approximation of Ampère's law,

$$H_1 l_1 + H_2 l_2 = Ni, \quad \text{we have}$$

$$\left(700 \frac{At}{m}\right)(1.3 m) + \left(600 \frac{At}{m}\right)(0.4m) = 500 i$$

Solving for i gives i = 2.3 A . Answer is b ●

MACHINERY 2

In the three-phase system below V_{AB} = 220 $\angle 0°$ V and Z = 8 + 8j ohms. Find I_A.

(a) 6.5 $\angle -30°$

(b) 19.5 $\angle -75°$

(c) 19.5 $\angle -45°$

(d) 33.8 $\angle -75°$

(e) 33.8 $\angle -15°$

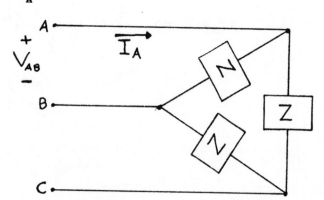

Solution

$$Z_\Delta = 8 + 8j \ \Omega = 11.3 \, \underline{/45°}$$

The load current is $\dfrac{V_{AB}}{Z_\Delta} = \dfrac{220 \, \underline{/0°}}{11.3 \, \underline{/45°}} = 19.5 \, \underline{/-45°}$

Then $I_A = \sqrt{3} \ \bar{I}_{AB} \, \underline{/-30°}$

$\qquad = \sqrt{3} \left(19.5 \, \underline{/-45°} \right) \left(\underline{/-30°} \right)$

$\qquad = 33.77 \, \underline{/-75°} \ A \bullet$ Answer is d.

MACHINERY 3

For the circuit in the previous problem, what is the total real power delivered to the load?

         (a)   5,240 W
         (b)   9,099 W
         (c)   15,720 W
         (d)   11,075 W
         (e)   6,055 W

Solution

From the previous problem $\quad V_{AB} = 220 \, \underline{/0°}$
$\qquad\qquad\qquad\qquad$ and $\qquad I_{AB} = 33.77 \, \underline{/-75°}$

Using the equation

$$P_T = \sqrt{3} \ V_L \, I_L \cos \theta$$

we have

$$P_T = \sqrt{3} \, (220)(33.77) \cos 45° = 9099 \ W \bullet$$

Note that θ is the angle between the phase quantities.

$\qquad\qquad$ Answer is b.

MACHINERY 4

A transformer is represented by the approximate equivalent circuit below. The power factor is .85 lagging. $I_z = 50$ A and $V_2 = 240$ V. Using V_2 as the reference $(240 \angle 0° V)$, find V_1.

(a) $2285 \angle -9.7°$ V

(b) $2285 \angle 41.5°$ V

(c) $1926 \angle -64.10°$ V

(d) $1926 \angle -32.3°$ V

(e) $2074 \angle -1.2°$ V

Solution

The power factor angle is θ, $\theta = -\cos^{-1}(.85) = -31.8°$

$E_2 = I_2 Z_2 + V_2 = (50 \angle -31.8)(6 \angle 10°) + 240 \angle 0°$
$= 300 \angle -21.8° + 240 = 518.55 - 111.4j$

For the ideal transformer $\dfrac{E_1}{E_2} = \dfrac{4}{1}$

So $E_1 = 4 E_2 = 2074.2 - j445.6$

I_2' can be found from I_2. $\dfrac{I_2'}{I_2} = \dfrac{1}{4}$

$I_2' = \dfrac{I_2}{4} = \dfrac{50 \angle -31.8°}{4} = 12.5 \angle -31.8°$

$\therefore V_1 = I_2' Z_1 + E_1 = (12.5 \angle -31.8°)(15 \angle 50°) + E_1$
$= 178.1 + j58.6 + 2074.2 - j445.6$
$= 2252.3 - j387.0$
$= 2285 \angle -9.7°$ ●

Answer is a.

MACHINERY 5

For the transformer in the previous problem, find I_1.

$\qquad$ (a) $\quad 12.5 \ \angle{-31.8°} \quad$ A

$\qquad$ (b) $\quad 12.5 \ \angle{0°} \quad$ A

$\qquad$ (c) $\quad 15.7 \ \angle{52°} \quad$ A

$\qquad$ (d) $\quad 15.7 \ \angle{-52°} \quad$ A

$\qquad$ (e) $\quad 18.4 \ \angle{-45°} \quad$ A

Solution

From the previous problem $I_2' = 12.5 \angle{-31.8°}$ and $V_1 = 2285 \angle{-9.7°}$.

$$I_1 = I_2' + V_1 Y_{\phi_1}$$
$$= 12.5 \ \angle{-31.8°} + (2285 \angle{-9.7°})(3 \ \angle{-60} \ m\mho)$$
$$= 10.6 - j6.6 + 6.855 \angle{-69.7°} = 13 - j13$$
$$= 18.4 \ \angle{-45°} \ A \ \bullet$$

Answer is e.

MACHINERY 6

Power is supplied to an industrial plant at 5000 V. The load is composed of 40 kW of heating (pf = 1) and 160 kVA of induction motors. The motors operate at .70 lagging power factor. Find the total current into the plant.

$\qquad$ (a) $\quad$ 42 A
$\qquad$ (b) $\quad$ 32 A
$\qquad$ (c) $\quad$ 38 A
$\qquad$ (d) $\quad$ 48 A
$\qquad$ (e) $\quad$ 30 A

Solution

For the motors $\theta = \cos^{-1}(0.70) = 45.6°$

The real and reactive powers are:

$P = 160 \cos 45.6° = 111.9 \text{ kW}$

$P_x = 160 \sin 45.6° = 114.3 \text{ kVAR}$

For the heating load $\theta = \cos^{-1}(1) = 0$

$P = 40 \text{ kW}$

$P_A = 40 \text{ kVA}$

The total plant current is

$$I = \frac{P_A}{V} = \frac{(40 \text{ kW} + 111.9 \text{ kW}) + j114.3 \text{ kVAR}}{5000 \text{ V}}$$

$$I = 38.0 \text{ A} \bullet$$

Answer is C.

MACHINERY 7

Given the ideal transformer, ac source, and load as shown. If the transformer and load were disconnected and, in their place, an equivalent load resistance replaced them, determine the equivalent resistance.

2:1

(a)   10.0 ohms
(b)   2.5 ohms
(c)   20.0 ohms
(d)   5.0 ohms
(e)   Unable to determine since the magnitude of V_s is not given.

Since the current on the left side of the transformer is 1/2 that of the right, and the voltage is twice the magnitude of the right side, then, since R = V/I, the equivalent resistance is given as the square of the turns ratio.   Thus

$R_{eq} = 4R_L = 4 \times 5 = 20$ ohms $\bullet$

Answer is (c)

10
CHEMISTRY

CHEMISTRY 1

All of the following are oxidation-reductions reactions, EXCEPT:

(a) $CaCO_3 \longrightarrow CaO + CO_2$

(b) $CO_2 + C \longrightarrow 2CO$

(c) $Fe + S \longrightarrow FeS$

(d) $2SO_2 + O_2 \longrightarrow 2SO_3$

(e) $Hg + I_2 \longrightarrow HgI_2$

Oxidation-reduction involves changes in oxidation state of the elements.

In (a): Ca remains at +2, C remains at +4, and O remains at -2. ●

(b): C is oxidized from 0 to +2 and is reduced from +4 to +2 while O remains at -2.

(c): Fe is oxidized from 0 to +2, while S is reduced from 0 to -2.

(d): S is oxidized from +4 to +6, while some of the O is reduced from 0 to -2.

(e): Hg is oxidized from 0 to +2, while I is reduced from 0 to -1.

Answer is (a)

CHEMISTRY 2

To determine the chemical formula of a compound, we need to know all of the following EXCEPT the

(a) elements that compose the compound.
(b) atomic weights of the elements in the compound.
(c) molecular weight of the compound.
(d) percentage composition of the compound.
(e) density of the compound.

All except density are required. ●

Example: A compound of nitrogen (at. wt. 14) and oxygen (at. wt. 16), of molecular wt. 92, contains 30.5 wt.% nitrogen. Determine its chemical formula.

Basis of calculation: 100 lb of compound

lb moles nitrogen $= \dfrac{30.5}{14} = 2.18$ lb moles oxygen $= \dfrac{69.5}{16} = 4.35$

Simplest ratio: NO_2 has molecular wt. of $14 + 32 = 46$

next higher ratio: N_2O_4 gives correct molecular wt. of 92.

Although the simplest formula is NO_2, the compound exists as the dimer, N_2O_4, nitrogen tetroxide. Other examples of multiples of the simplest formula are P_4 and Hg_2Cl_2.

We note that vapor density may be used to determine molecular weight of gases, while solid density is not directly applicable.

Answer is (e)

CHEMISTRY 3

The number of molecules in 22.4 liters (under standard conditions) of a substance in its gaseous state, is called

    (a) Dulong's number
    (b) Petit's number
    (c) Avogadro's number
    (d) Gay-Lussac's number
    (e) Graham's number

One mole of any gas under standard conditions occupies 22.4 liters. Each mole of gas has a stated number of molecules. This value, 6.02×10^{23}, is known as the Avogadro number. ●

Answer is (c)

CHEMISTRY 4

Which of the following metals is the best heat conductor?

    (a) aluminum
    (b) copper
    (c) gold
    (d) silver
    (e) 1020 steel

In metals there is a relationship between thermal and electrical conductivity. Resistivity is the reciprocal of conductivity.

	Thermal Conductivity, k BTU/hr $ft^2(^\circ F/in)$	Electrical Resistivity, ρ ohms(cir mil)/ft
Silver	2900 ●	9.8
Copper	2700	10.3
Gold	2060	14.7
Aluminum	1530	18.5
1020 steel	360	60.

(Gold is used in microelectronics because it does not tarnish.)

Answer is (d)

CHEMISTRY 5

One equivalent weight of H_2SO_4 is equal to

  (a) 98.06 g
  (b)   2 g
  (c) 49.03 g
  (d) 96.06 g
  (e) 32.06 g

  Atomic wts: H = 1.00
  S = 32.06
  O = 16.00

━━━━━

One equivalent weight of $H_2SO_4 = \dfrac{2(1) + 32.06 + 4(16)}{2} = 49.03$ g ●

Answer is (c)

CHEMISTRY 6

When exposed to the atmosphere, which of the following <u>liquids</u> is coldest?

  (a) oxygen
  (b) argon
  (c) nitrogen
  (d) hydrogen
  (e) helium

━━━━━

Liquid	Atmospheric Boiling Point
oxygen	-183°C
argon	-189°
nitrogen	-196°
hydrogen	-253°
helium	-269° ●

Answer is (e)

CHEMISTRY 7

Hard water is water which contains soluble salts of which of the following elements:

  (a) Sodium
  (b) Sulfur
  (c) Calcium
  (d) Chlorine
  (e) Nitrogen

━━━━━

Soluble salts of calcium and magnesium are generally associated with hardness. The degree of hardness in water, no matter what the cause, is usually expressed in terms of parts per million of calcium carbonate. ●

Answer is (c)

CHEMISTRY 8

Which of the following metals has the highest specific heat capacity at 100°C?

(a) aluminum
(b) bismuth
(c) copper
(d) iron
(e) silver

Dulong & Petit Law

$$\text{atomic } C_p \text{ of metals} \approx \frac{6.4 \text{ BTU/}^\circ\text{F}}{\text{lb. mole}}$$

Thus the metal with the lowest atomic weight will have the highest specific heat capacity. Of the group of metals listed, aluminum has the lowest atomic weight. It therefore has the highest specific heat capacity. ●

Answer is (a)

CHEMISTRY 9

Which of the following is a member of the halogen family?

(a) sodium
(b) fluorine
(c) hydrogen chloride
(d) phosphorus
(e) hydrogen

The halogen family ("salt producing") consists of fluorine, chlorine, bromine, and iodine. ●

Answer is (b)

CHEMISTRY 10

A sample of a substance, to which a chemical formula can be assigned, whose weight is equal to its formula weight is termed a:

(a) molecule
(b) mole
(c) gram equivalent weight
(d) one normal solution
(e) atom

Answer is (b) ●

CHEMISTRY 11

The radical ion $\left(NO_2\right)^-$ is called a

(a) nitrate
(b) nitrite
(c) nitride
(d) binitrite
(e) binitrate

Answer is (b) ●

CHEMISTRY 12

Hydrogen is common to all

(a) acids
(b) salts
(c) oxides
(d) metals
(e) inorganic material

Answer is (a) ●

CHEMISTRY 13

Which of the following does not illustrate the effect of temperature or pressure on gas solubility?

(a) Air bubbles form on the sides of a warm glass of water.
(b) Soda pop is bottled under pressure.
(c) Boiling frees water of gases.
(d) Air is more humid on rainy days.
(e) Relative humidity is dependent on temperature.

Answer is (d) ●

CHEMISTRY 14

The chemical process which occurs when water is added to cement is

(a) oxidation
(b) Brownian movement
(c) plastic flow
(d) hydration
(e) counter diffusion

Answer is (d) ●

CHEMISTRY 15

An amphoteric hydroxide is one which:

(a) has a valence of -2.
(b) reacts violently with water.
(c) has been created by bombardment with high energy protons.
(d) may act as either a base or an acid in chemical reactions.
(e) decays spontaneously.

Answer is (d) ●

CHEMISTRY 16

All statements about the equilibrium reactions below are correct, EXCEPT:

(a) $N_2 + 3H_2 \rightleftharpoons 2NH_3$ is shifted to the right by pressure increase.

(b) $CaCO_3 \rightleftharpoons CaO + CO_2$ is shifted to the left by pressure increase.

(c) $Ca^{++} + SO_4^{--} \rightleftharpoons CaSO_4$ is shifted to the right by precipitating $CaSO_4$.

(d) $Ag^+ + 2NH_3 \rightleftharpoons Ag(NH_3)_2^+$ is shifted to the right by raising concentration of Ag^+.

(e) $N_2O_4 \rightleftharpoons 2NO_2$ is shifted to the right by pressure increase.

LeChatelier's principle: Systems respond to reduce applied stress. This principle is used to qualitatively estimate the effect of altered solution concentrations or pressures on chemical equilibria. Increasing reactant concentration produces more product (to reduce reactant concentration); raising pressure favors fewer gas molecules (to reduce pressure); and vice versa.

The false statement is (e) ●

CHEMISTRY 17

All of the following statements about pH of a solution are correct, EXCEPT:

(a) It is $-\log_{10}\left[H^+\right]$

(b) It is held reasonably constant in buffer solutions.

(c) pH 10 is ten times more alkaline than pH 9.

(d) pH of tap water is below 7 due to dissolved CO_2.

(e) It is zero for neutral solutions.

pH is $-\log_{10}\left[H^+\right]$ Hydrogen ion concentration is gram moles/liter over the range from 1 to 10^{-14} is expressed by pH from 0 to 14. Buffer solution contain salts of weak acids or weak bases; they stabilize pH of solutions despite dilution or small additions of acid or base. pH of acid solutions is below 7; of neutral solutions is equal to 7; and of alkaline solutions is above 7.

The false statement is (e) ●

11
STRUCTURE
OF MATTER

STRUCTURE OF MATTER 1

The element tin has ten different stable isotopes. The atomic nuclei of all isotopes have the same:

        (a)   number of neutrons.
        (b)   number of protons.
        (c)   radius.
        (d)   mass.
        (e)   binding energy.

Nuclei of the various isotopes of tin have the same atomic number Z = 50 protons but different mass numbers, A, ranging from 112 to 124 because of differing A-Z number of neutrons. Nuclear radii and configurations differ. Binding energy is derived from the mass discrepancy between mass of the whole nucleus and summation of masses of its individual nucleons (protons and neutrons). One atomic mass unit (amu) equals 931 Mev by the $E = mc^2$ energy-mass equivalence, and one amu is 1/12 the mass of an atom of $_6C^{12}$ (the basis of atomic weights) or 1.66×10^{-24} gram. Binding energy <u>per</u> <u>nucleon</u> varies with mass number, averaging 8 Mev/nucleon, but being much less for the lightest, and slightly less for the heaviest nuclei. Energy release and stability gained through thermonuclear fusion of light elements and fission of the heaviest elements are consequences of binding energy differences.

Answer is (b) ●

STRUCTURE OF MATTER 2

All of the following statements about the electron structure of atoms are correct, EXCEPT:

    (a)   Electrons of highest energy are located in the outer orbitals.

    (b)   Each orbital can contain one electron.

    (c)   K, L, M, etc. shells are designations representing the principal quantum numbers n = 1, 2, 3 etc.

    (d)   Each shell may contain up to n^2 orbitals, but not exceeding 16.

    (e)   One s, three p, five d and seven f orbitals, containing up to 2, 6, 10 and 14 electrons respectively, have different spatial configurations.

Four types of quantum numbers describe the building up of shells of electron orbitals with discretely increasing energy level for each orbital.

<u>Principal quantum number</u>, n, represents a major grouping of energy levels from 1 to 7, sometimes referred to as K, L, M etc. shells.

<u>Azimuthal quantum number</u>, ℓ, denotes different shapes of orbitals within the major groupings. These shapes are labeled s, p, d and f for ℓ = 0, 1, 2 and 3 respectively.

<u>Magnetic quantum number</u>, m, may be +1, -1 or 0. Each value of m defines a magnetic property of the electron resulting from its orbital motion.

<u>Spin magnetic quantum number</u>, s, may have a value of +1/2 or -1/2 depending on direction of electron spin. This permits each orbital to contain two electrons of opposite spin.

One s orbital, three p orbitals, five d orbitals and seven f orbitals exist. Each orbital can contain two electrons. The maximum number of electrons in each major energy level (n = 1, 2, 3, ...) is $2n^2$, but not exceeding 32.

These numbers identify an overlapping succession of electron orbitals designated, in order of increasing energy: 1s, 2s, 2p, 3s, 3p, 4s, 3d, 4p, 5s, 4d, 5p, 6s, 4f, 5d, ...

Note that s orbitals exist at n = 1 upward, p orbitals at n = 2 upward, d orbitals at n = 3 upward, etc.

Energy levels are filled sequentially, beginning with the lowest 1s level. In the ground state the lowest energy levels are filled and no vacancies exist.

Exited states result when an electron receives energy, vacates a lower energy orbital, and occupies a higher energy orbital.

One photon of discrete energy is emitted or absorbed as an electron makes transition between differing energy levels of the various orbitals.

The false statement is (b) ●

STRUCTURE OF MATTER 3

All of the following elementary particles have detectable mass, EXCEPT:

    (a)   neutron n^0

    (b)   proton p^+

    (c)   electron e^-

    (d)   positron e^+

    (e)   neutrino ν

The field of high energy particle physics has shown the existence of more than 30 elementary particles, including all of the above plus various mesons, hyperons and other particles. The neutrino is a massless, unchanged particle that can possess both energy and momentum; it is emitted together with an electron during beta decay. The proton p^+ (and antiproton p^-), electron e^- (and positron e^+) and neutrino ν (and antineutrino $\bar{\nu}$) are stable outside the nucleus and do not decay prior to reaction; the neutron n^0 has a half life of 12 minutes. All the others have lifetimes much less than 10^{-6} second and decay spontaneously into other particles, react with other particles, or are annihilated with energy release on combination with their antiparticles.

Answer is (e) ●

STRUCTURE OF MATTER 4

Which of the following modes of radioactive decay results in an increased Z or number of protons in the nucleus?

(a) α decay (helium nucleus emission)

(b) negative beta decay (electron emission)

(c) positive beta decay (positron emission)

(d) K electron capture (unchanged neutrino emission)

(e) γ emission (photon energy emission)

〰〰〰〰〰

Radioactive decay is summarized below:

Radioactive decay mode	net N = A-Z, change in no. of neutrons	net Z, change in no. of protons	Reaction within nucleus
α decay	-2	-2	emission of $(_2He^4)^{++}$ reduces nucleus by 2 neutrons and 2 protons
negative beta decay	-1	+1	$n^o \longrightarrow p^+ + e^- \nearrow + \bar{\nu} \nearrow$ electron and antineutrino emitted
positive beta decay	+1	-1	$p^+ \longrightarrow n^o + e^+ \nearrow + \nu \nearrow$ positron and neutrino emitted
electron capture	+1	-1	$p^+ + e^- \longrightarrow n^o + \nu \nearrow$ neutrino emitted
γ emission	0	0	excited nucleus loses excess energy by energetic photon emission

Radioactive nuclei spontaneously decay and transform themselves into other nuclear species with a more stable neutron-proton ratio and configuration. Often a succession of α - and β -decays, with accompanying γ -emission, are required before a nucleus reaches stability. In electron capture the nucleus captures an inner K orbital electron and emits a neutrino, accompanied by x rays due to electronic shifts characteristic of the new atom. Electron capture and positive beta decay (positron emission) yield the same product, though positron emission is rare and does not occur in the naturally radioactive isotopes.

If $_{53}I^{131}$ could hypothetically decay by each of the listed modes the daughter product formed and radiation emitted would be:

$$\alpha \text{ decay:} \qquad _{51}Sb^{127} + _2He^4$$

$$\text{negative beta:} \qquad _{54}Xe^{131} + e^- + \bar{\nu}$$

$$\text{positive beta:} \quad _{52}\text{Te}^{131} + e^+ + \nu$$

$$\text{electron capture:} \quad _{52}\text{Te}^{131} + \nu$$

$$\gamma \text{ emission:} \quad _{53}\text{I}^{131} + \gamma$$

(The actual decay mode is mostly by negative beta decay to stable $_{54}\text{Xe}^{131}$.)

Answer is (b) ●

STRUCTURE OF MATTER 5

$_{38}\text{Sr}^{90}$ decays by negative beta emission and has a halflife of 28 years. All of following statements are correct, EXCEPT:

(a) Mean lifetime of $_{38}\text{Sr}^{90}$ is 40 years.

(b) Activity of 10 micrograms of $_{38}\text{Sr}^{90}$ is 1.4 millicuries.

(c) First daughter product of the beta decay is $_{39}\text{Y}^{90}$.

(d) Decay probability per year for any atom of $_{38}\text{Sr}^{90}$ is numerically equal to the decay constant of 0.025/year.

(e) After 100 years more than 10% of the original $_{38}\text{Sr}^{90}$ remains.

~~~~~~~~~~~~

Radioactive decay is statistical in nature and follows an exponential law. Fraction remaining is equal to $e^{-\lambda t}$ where λ is the decay constant.

Halflife $t_{\frac{1}{2}}$ is related to decay constant by $\lambda = \dfrac{\ln 2}{t_{\frac{1}{2}}} = \dfrac{0.693}{t_{\frac{1}{2}}}$

$$\lambda = \frac{0.693}{28} = 0.025/\text{year}.$$

Mean lifetime of all of the atoms, $t = \dfrac{1}{\lambda} = 1.44 t_{\frac{1}{2}}$,

thus mean lifetime of $_{38}\text{Sr}^{90}$ is 1.44(28) = 40. years.

Activity, number of disintegrations/sec, is expressed in curies. One curie equals 3.70×10^{10} disintegrations/sec.

Activity = λN where λ is decay constant/sec and N is number of atoms of isotope present.

$$\lambda = \frac{0.025}{\text{year}} \times \frac{\text{year}}{365 \text{ days}} \times \frac{\text{day}}{24 \text{ hrs}} \times \frac{\text{hour}}{3600 \text{ sec}} = 7.9 \times 10^{-10}/\text{sec}.$$

$$N = \frac{10 \times 10^{-6} \text{ grams}}{90 \text{ grams/gm atom}} \times 6.025 \times 10^{23} \text{ atoms/gm atom} = 6.7 \times 10^{16} \text{ atoms present}$$

Activity = λN = $(7.9 \times 10^{-10}/\text{sec})(6.7 \times 10^{16} \text{ atoms})$ = 5.3×10^7 disintegrations per second

$$= \frac{5.3 \times 10^7}{3.70 \times 10^{10}} = 1.4 \times 10^{-3} \text{ curies or 1.4 millicuries}.$$

Thus, short halflives yield more activity, disintegrations/sec.

Since negative beta decay converts one neutron to a proton and ejects an electron, the atomic number increases by one to 39, the name changes accordingly from 38 = Strontium to 39 = Yttrium, and the mass number, or total neutrons plus protons, remains unchanged at 90.

Result: first daughter product is $_{39}Y^{90}$.

Decay constant is almost exactly the probability that any one atom will decay in the time interval. One half of the original amount remains after one halflife, $\frac{1}{4}$ after two halflives, $\frac{1}{8}$ after three halflives, and a trivial amount remains forever. After 100 years, the fraction remaining is $e^{-.025(100)} = e^{-2.5} = 0.082$

The false statement is (e) ●

STRUCTURE OF MATTER 6

Radioactive $_6C^{14}$ decays at a rate proportional to the amount present at any time. Its halflife is 5600 years. If the amount of $_6C^{14}$ present in wood charcoal taken from a ruin is 40% of the amount in timber growing now, estimated age of the charcoal is nearest to:

    (a)   4200 years
    (b)   5600 years
    (c)   7400 years
    (d)   8400 years
    (e)   9500 years

Decay constant λ is related to halflife by

$$\lambda = \frac{\ln 2}{t_{\frac{1}{2}}} = \frac{0.693}{t_{\frac{1}{2}}} = \frac{0.693}{5600} = 1.24 \times 10^{-4}/\text{year}.$$

Alternately, the decay constant λ can be stated as a disintegration rate of one part in about 8000 per year.

Fraction present $= e^{-\lambda t}$, where t is time elapsed.

$$0.40 = e^{-1.24 \times 10^{-4}t} \qquad 1.24 \times 10^{-4}t = \text{antiln}_e\ 0.40 = 0.915$$

$$t = \frac{0.915}{1.24 \times 10^{-4}} = 7400 \text{ years} \quad ●$$

Answer is (c)

STRUCTURE OF MATTER 7

Which of the following nuclear reactions is thermonuclear?

    (a) $_7N^{14} + _0n^1 \longrightarrow _6C^{14} + _1H^1$

    (b) $_5B^{10} + _0n^1 \longrightarrow _3Li^7 + _2He^4$

    (c) $_{92}U^{238} + _0n^1 \longrightarrow _{92}U^{239} + \gamma$

    (d) $_1H^3 + _1H^2 \longrightarrow _2He^4 + _0n^1$

    (e) $_{92}U^{235} + _0n^1 \longrightarrow 2 \text{ fission fragments} + 1 \text{ to } 3\ _0n^1$

〰〰〰〰〰〰〰

Nomenclature for nuclear reactions uses $_Z(\text{name})^A$ for nuclei and particles.

Z is atomic number corresponding to name, and A is mass number.

Thus: proton (p^+) is shown as $_1H^1$, neutron (n^0) as $_0n^1$, and the alpha

particle (α) as $_2He^4$.

Superscripts and subscripts must balance for all valid reactions. Although all nuclear reactions involve conservation of mass-energy, electrical charge and number of nucleons, skeleton reactions are usually written as above. Type reactions may be further abbreviated showing the input-output particles or emissions:

n - p   for neutron absorption and proton emission

n - α for neutron absorption and α emission

n - γ for neutron adsorption and γ emission

n - f   for neutron adsorption and fission

(a) is an n - p reaction that produces $_6C^{14}$ in the upper atmosphere when neutrons from cosmic ray events bombard atmospheric nitrogen.

(b) is an n - α reaction used both for detection of thermal neutrons and for reactor control.

(c) is an n - γ neutron capture reaction using abundant $_{92}U^{238}$ to produce $_{92}U^{239}$ which decays to $_{94}Pu^{239}$ in a few days.

(d) is the triton - deuteron thermonuclear reaction which can be maintained only at extremely high temperature and pressure. Energy release is 17.6 Mev per event. Similar reactions involving fusion of light nuclei into heavier ones occur in the sun and stars. ●

(e) is an n - f fission reaction that produces two distributions of fission fragments in the ranges of 84 - 104 and 130 - 149 with maxima at 93 and 134. It is accompanied by a competing n - γ reaction that can consume more than half of the average 2.5 neutrons released per event.

Answer is (d)

STRUCTURE OF MATTER 8

Which of the following isotopes would be expected as product of an α - n reaction on $_4Be^9$?

(a) $_5B^9$ (d) $_2He^6$

(b) $_4Be^{10}$ (e) $_5B^{12}$

(c) $_6C^{12}$

〰〰〰〰〰〰〰

(a) results from a p - n reaction, (b) from n - γ, (c) from α- n
(d) from n - α , and (e) from α- p.

$$_4Be^9 + {}_2He^4 \longrightarrow {}_6C^{12} + {}_0n^1 \quad \bullet$$

Answer is (c)

STRUCTURE OF MATTER 9

All of the following statements about cross section, σ, are correct, EXCEPT:

   (a) Cross section is expressed in barns (10^{-24} cm^2).
   (b) Cross section indicates the likelihood of interaction between an incident particle and target nucleus.
   (c) One barn is of the order of magnitude of the geometrical cross section of the nucleus.
   (d) Maximum cross section is about 100 barns.
   (e) Cross sections vary with type reaction and energy of incident particles, often having peaks at specific energies.

~~~~~~~~~~~~~~~

Cross section σ may be subscripted:

σ_a indicating absorption (capture, or n - γ reaction),

σ_s for scattering,

σ_f for fission,

and for various other reactions, such as

σ_p for n - p reaction.

Competing reactions each have their own cross section which varies widely with incident particle energy. Examples for thermal energy neutrons:

(1) $_{92}U^{235}$ has σ = 683 for n - γ capture, and σ_f = 577 for fission.

(2) $_{48}Cd^{113}$ has σ = 2.1 x 10^4 and is used for reactor control.

(3) $_8O^{16}$ has σ = 2 x 10^{-5}, hence is very unreactive with thermal neutrons.

(4) $_1H^2$ deuterium (D) has σ = 5.7 x 10^{-4}, $_6C^{12}$ has σ = 3.3 x 10^{-3} while $_1H^1$ hydrogen has σ = 0.33

Therefore heavy water D$_2$O is an excellent moderator for slowing fast neutrons to thermal energy in reactors because energy is dissipated on collision and reaction seldom occurs.

The false statement is (d) $\bullet$

STRUCTURE OF MATTER 10

All of the following are characteristic of waves, EXCEPT:

- (a) Energy of a wave varies with the square of its amplitude.
- (b) Interference is interaction of waves. The observed effect is based on the principle of superposition.
- (c) Diffraction, or bending of a wave, occurs around the edges of obstacles.
- (d) Refraction, or bending of a beam, occurs when it passes between media with different velocities of propagation.
- (e) Wave frequency changes on entering a medium where the velocity of propagation is changed.

~~~~~~~~~~~~~~~~

Energy may be dissipated by absorption (attenuation) during passage through a medium. Conservation of energy requires that waves radiating outward from the source must deliver lesser amounts of energy per unit area since the receiving area increases with distance squared. Energy contained in most waves varies with the square of the amplitude of displacement from equilibrium.

Refraction, reflection, interference, and diffraction are characteristic of all wave motion. Diffraction effects become evident when obstacle dimensions are comparable with wavelength.

Frequency remains unchanged in passing between media. ●

<div align="center">The false statement is (e)</div>

STRUCTURE OF MATTER 11

All of the statements about electromagnetic waves are correct, EXCEPT:

- (a) Electromagnetic waves are transverse waves that consist of mutually coupled, changing electric and magnetic fields.
- (b) Directions of the electric field, of the magnetic field, and of propagation are mutually perpendicular.
- (c) Field intensity changes take place, but no displacement of mass occurs in the path of electromagnetic waves.
- (d) Electric field maxima and minima occur out of phase with those of the associated magnetic field.
- (e) Velocity of propagation is dependent only on permittivity, ϵ , and permeability, μ , of the medium.

~~~~~~~~~~~~~~~~

Polarization proves that electromagnetic waves are transverse, for longitudinal waves cannot be polarized. Unlike electromagnetic waves, other types of waves cause displacement of mass. Electromagnetic waves are propagated in vacuum and in dielectric materials, but are attenuated due to I^2R losses in conductors. Electric field maxima and minima occur simultaneously and <u>in phase</u> (except very close to the oscillator).

Velocity of propagation $v = \sqrt{1/\epsilon\mu}$ in materials is less than speed of light in vacuum, $c = 3 \times 10^8$ meters/sec.

The false statement is (d) ●

STRUCTURE OF MATTER 12

All of the following statements about sound are correct, EXCEPT:

    (a) Sound is propagated by longitudinal compressional waves.

    (b) Velocity of sound in solids may be calculated from their elastic modulus and density.

    (c) Velocity of sound in gases is independent of pressure.

    (d) Sound waves do not exhibit refraction in passing between media.

    (e) Velocity of wave propagation equals the product of frequency and wavelength.

Velocity of longitudinal sound waves in solids may be calculated from elastic modulus and density: v in ft/sec $= \sqrt{\dfrac{E\, g_c}{\rho\, 12}}$ where E is tensile-conpressional (Youngs) modulus in psi, ρ is density in lb/cu in, with $g_c = 32.2$ and 12 in/ft being inserted for correct dimensions. Shear modulus G replaces E for transverse waves. Bulk modulus of elasticity K replaces E for liquids. In gases $v = \sqrt{k\, g_c\, R\, T}$ where $k = C_p/C_v$, R is gas constant in ft $lb_f/lb_m \,^{\circ}R$, and T is absolute temperature in $^{\circ}R$.

Approximate velocity of sound at ambient temperature is 1130 ft/sec in air, 4720 ft/sec in water and 15,800 ft/sec in steel.

Both longitudinal and electromagnetic waves exhibit reflection, refraction, diffraction, and interference. Sound waves are refracted (bent) in passing between media in a manner analogous to light waves; Snells law applies. Index of refraction is the ratio of velocities in the two media. Frequency remains constant and wavelength varies because $v = f \lambda$. ●

The false statement is (d)

STRUCTURE OF MATTER 13

All of the following statements about sound intensity are correct, EXCEPT:

    (a) Sound intensity level is energy expressed in decibels,

$$db = 10\ \log_{10}\left(\frac{I}{I_o}\right)$$

    (b) Zero level of sound intensity, I_o, is established at an energy of 10^{-16} watt/cm^2.

    (c) Two sound sources, each individually of 70 db sound intensity, together produce an intensity of 80 db.

    (d) Intensity of the lower audible limit at 50 cycles/sec is approximately 10^6 times (i.e., 60 db above) that existent at 3000 cycles/sec.

    (e) The human ear is most sensitive to sound at 3000 cycles/sec, and is responsive in young adults from about 20 to 20,000 cycles/sec.

Sound intensity or actual energy per unit area is expressed on the logarithmic decibel scale with the zero level at 10^{-16} watt/cm^2. One sound source of 70 db has an energy 10^7 times as intense as reference, two identical 70 db sound sources have energy 2×10^7 times as intense as reference, or 73 db above reference. One 70 db source plus one 75 db source have combined energies of

1×10^7 plus 3.16×10^7 or 4.16×10^7 times reference.

Since $\log_{10}$ of 4.16×10^7 is 7.62, the combined energy is 76.2 db above reference.

Loudness of noise, being dependent upon the frequency response and sensitivity of the human ear, also is expressed on scales of loudness level in phons, and of loudness sensation in millisones.

The false statement is (c) ●

STRUCTURE OF MATTER 14

At which frequency in cycles/sec does an observer hear a 1000 cycle/sec tone if both source and observer are each moving towards each other at 88 ft/sec? Sound velocity is 1100 ft/sec in air.

(a)   852
(b)   926
(c)   1080
(d)   1088
(e)   1174

The Doppler effect is an apparent frequency shift experienced by the observer due to relative motions. This applies to both light and sound, and explains the spectral red shift observed astronomically.

A moving source compresses or expands relative wavelength experienced by the observer, and a moving observer increases or decreases relative propagation velocity of the wave train. $v = f \lambda$ applies.

	Relative v in ft/sec observed due to motion of observer	Relative λ in ft observed due to motion of source	$f = v/\lambda$ frequency observed in cycles/sec
Stationary source Moving observer $S \; \leftarrow O$ $S \qquad O \rightarrow$	increased or decreased v $1100 + 88 = 1188$ ft/sec $1100 - 88 = 1012$ ft/sec	constant λ $\frac{1100}{1000} = 1.100$ 1.100	$\frac{1188}{1.100} = 1080$ 920
Moving source Stationary observer $S \rightarrow \quad O$ $\leftarrow S \quad O$	constant v 1100 1100	compressed or expanded λ $1.100(1012/1100) = 1.012$ $1.100(1188/1100) = 1.118$	1088 926
Moving source Moving observer $S \rightarrow \leftarrow O$ $S \rightarrow \quad O \rightarrow$ $\leftarrow S \quad \leftarrow O$ $\leftarrow S \quad O \rightarrow$	increased or decreased v 1188 1012 1188 1012	compressed or expanded λ 1.012 1.012 1.188 1.188	1174 ● 1000 1000 852

Answer is (e)

STRUCTURE OF MATTER 15

Frequency f of monochromatic yellow light which has a wavelength λ of 589 nm (nanometers) in vacuum is nearest to:

(a)   295,000 GHz
(b)   510,000 GHz
(c)   589,000 GHz
(d)   1,020,000 GHz
(e)   1,180,000 GHz

kHz $= 10^3$ Hz $= 10^3$ cycles/sec,   MHz $= 10^6$ Hz,   GHz $= 10^9$ Hz.

nanometer $= 10^{-9}$ meter $=$ millimicron $= 10$ Å

$= 589$ nm $= 589 \times 10^{-9}$ m $= 5890$ Å,   v $= f\lambda$ $= 3.00 \times 10^8$ m/sec

3.00×10^8 m/sec $= f(589 \times 10^{-9}$m$)$

$$f = \frac{3.00 \times 10^8}{589 \times 10^{-9}} = 5.1 \times 10^{14} \text{ cycles/sec} = 510{,}000 \text{ GHz} \; \bullet$$

Answer is (b)

STRUCTURE OF MATTER 16

A radio station broadcasts at a frequency of 600 kilohertz. The corresponding wave length is closest to which of the following values?
Assume the velocity of sound is 300 meters/sec and the velocity of light is 3×10^8 meters/sec.

(a)   500 meters

(b) 2×10^{-3} meters

(c) 5×10^{-4} meters

(d)   2000 meters

(e)   50 meters

Speed of a wave $=$ Frequency x Wavelength   v $= f\lambda$

$$\lambda = \frac{v}{f} = \frac{3 \times 10^8 \text{ meters/sec}}{600{,}000/\text{sec}} = 500 \text{ meters} \; \bullet$$

Answer is (a)

STRUCTURE OF MATTER 17

All of the following statements about photons are correct, EXCEPT:

(a)   Photons are small quanta of electromagnetic energy.

(b)   Energy of photons varies according to E $=$ hf, where h is Planck's constant, and f is frequency.

(c)   Photon energy varies with source intensity, i.e. a bright light source emits photons of higher energy than a dim source.

(d)   Velocity of photons in vacuum is c, the speed of light, 3×10^{10} cm/sec.

(e)   Two light sources of the same wavelength emit photons of the same energy.

The discrete energy per photon is independent of intensity or number of photons emitted.

E $=$ hf where h $= 6.625 \times 10^{-34}$ joule-sec.

Velocity v, frequency f, and wavelength λ are related

by v $= f\lambda$.

The false statement is (c) $\bullet$

STRUCTURE OF MATTER 18

A ray of light passes from air into water. The angle of incidence is
30 degrees from the vertical. Assume the velocity of light in air and
vacuum to be 3×10^{10} cm/sec and in water to be 2.26×10^{10} cm/sec.
Find the angle of refraction relative to the vertical.

A ray of light undergoes an abrupt change of direction when passing obliquely
into another medium in which it travels at a different velocity. The ray
will be deviated toward the normal (vertical in the illustration) when the
velocity is reduced (the situation in this problem) and deviated away from
the normal when the velocity is increased.

The law of refraction:
When a wave travels obliquely from one medium into another, the ratio of
the sine of the angle of incidence to the sine of the angle of refraction
is the same as the ratio of the respective wave velocities in these media,
and is a constant for two particular media.

Thus:

$$\frac{\sin i}{\sin r} = \frac{v_1}{v_2} \qquad \sin r = \frac{v_2 \sin i}{v_1} = \frac{2.26 \times 10^{10} \times 0.5}{3 \times 10^{10}} = 0.3767$$

$$\text{Angle of Refraction (r)} = 22.1 \text{ degrees} \bullet$$

Total internal reflection can occur when an internal ray is reflected at
an interface. This occurs when incident angle exceeds a critical angle c,
determined from sin c = n of new media/n of old media, and is limited
to situations where new media is less dense (has lower n) than old media.
Example: a ray in glass may be reflected rather than penetrating a
          glass-air interface.

STRUCTURE OF MATTER 19

A one-candlepower light source radiates luminous flux at the rate of

    (a)   1 lumen
    (b)   2 lumens
    (c) π lumens
    (d) 2π lumens
    (e) 4π lumens

A one-candlepower source radiates 1 lumen upon a surface of 1 square foot at a
distance of 1 foot. If one imagines a spherical shell of 1 foot radius around
the 1 candlepower source, it is clear that each square foot receives 1 lumen.
The area of the spherical shell is $4\pi R^2$, therefore the 1 candlepower source
radiates 4π lumens. $\bullet$

Answer is (e)

12
COMPUTER PROGRAMMING

PROGRAMMING 1

Based on the FORTRAN statements, how many input data lines will be read?

```
    READ(60,10)L,(X(I),I=1,13)
 10 FORMAT(I5/(6F13.2))
```

(a) 1
(b) 2
(c) 3
(d) 4
(e) 14

[Please note that throughout this chapter the input device is assigned logical unit number 60 and the output device 61. This will vary from computer center to computer center. Often, for example, 5 and 6 are used. Otherwise, standard FORTRAN 77 is assumed.]

In this problem one value will be read from the first data line, and six values from subsequent data lines until the read statement is satisfied. It will therefore take four data lines to read the 14 values. ●

Answer is (d)

PROGRAMMING 2

How many data lines will be read by the following?

```
    DIMENSION L(43)
    READ(60,15)B,(L(I),I=1,33)
 15 FORMAT(F10.0/(5I5))
```

(a)   1
(b)   2
(c)   7
(d)   8
(e) 34

The value of B will be read from the first data line. Five values of L will be read from each subsequent data line until the read statement is satisfied. ●

Answer is (d)

PROGRAMMING 3

Consider the following

```
    READ(60,12)N
 12 FORMAT(I3)
    M=1
    DO 10 I=2,N
 10 M=M*I
    WRITE(61,12)M
    END
```

What is the relationship between input N and the output M?

(a) M = N

(b) M = $2N^2$

(c) M = N^{N-1}

(d) M = N!

(e) M = N^N

Answer is (d) ●

PROGRAMMING 4

Consider the following

```
    A=14.5
 15 IF(A.GT.10.)GO TO 30
 20 GO TO 25
 30 GO TO 35
```

The label of the next statement to be executed following this portion of the program is

(a)   15
(b)   20
(c)   25
(d)   30
(e)   35

This is the one branch logical IF statement. If the logical expression within the parentheses is *TRUE,* the statement to the right will be executed; otherwise it will not be executed.

In this problem 14.5 is Greater Than 10, hence the expression is true and one will go to the statement labelled 30 and then to statement 35. ●

Answer is (e)

PROGRAMMING 5

Consider the following

```
ID=3**2/3.3+1
GO TO(10,11,12,13,14),ID
```

The label of the next statement to be executed is

        (a)   10
        (b)   11
        (c)   12
        (d)   13
        (e)   14

The hierarchy of arithmetic operations is

    1. Exponentiation
    2. Multiplication and Division
    3. Addition and Substraction

The order of operations is from group 1 to group 2 to group 3. Within any group evaluation proceeds from left to right. Parentheses may be used to override this order of operations.

ID = $3^2/3.3$ +1 = 2.73 + 1 = 3.73 which truncates to 3

The next statement to be executed is 12 ●

Answer is (c)

PROGRAMMING 6

Consider the following

```
J=2*3**2/3.3-1.2*3+1
GO TO(5,4,3,6,2),J
```

What is the label of the next statement to be executed?

        (a)   2
        (b)   3
        (c)   4
        (d)   5
        (e)   6

Mathematically the computation is

$[2(3^2)]/3.3$ -(1.2*3) +1 = 2.85 which truncates to 2

Answer is (c) ●

PROGRAMMING 7

In the following statements select the one which is not standard FORTRAN.

    (a) GO TO(1,2,3,2),N

    (b) DO 50,I=1,5

    (c) GO TO 15

    (d) WRITE(61,20)

    (e) C(I)=A(I+1)**B(I)

The comma after 50 in (b) is not standard FORTRAN. The statement is acceptable, however, for use with some FORTRAN compilers.
Item (d) has a blank list. This often occurs when printing output headings, etc.

Answer is (b)

PROGRAMMING 8

In the following FORTRAN statements select the one that is incorrectly written.

    (a) X = 2,412.62

    (b) A = 412

    (c) IF(3.*A.EQ.B) GO TO 20

    (d) B=A**I

    (e) IF(X)10,20,10

A comma must not be used in numbers like 2412.62

Answer is (a)

PROGRAMMING 9

Which one of the following is a FORTRAN executable statement?

    (a) REAL
    (b) INTEGER
    (c) COMMON
    (d) EQUIVALENCE
    (e) END

The END statement signals the compiler the physical end of either a subprogram or the main program. The other four items are nonexecutable statements (declarations).

Answer is (e)

PROGRAMMING 10

(This is a 4-part problem)

Consider the following FORTRAN program.

```
Line
No.
 1       PROGRAM SORT
 2       DIMENSION LIST (100)
 3       READ(60,10) N
 4    10 FORMAT(I3)
 5       READ(60,15) (LIST(I), I=1,N)
 6    15 FORMAT(5X,10I4)
 7    20 DO 25 K=2,N
 8       J=K-1
 9       IF(LIST(K).GE.LIST(J)) GO TO 25
10       NTEMP = LIST(J)
11       LIST(J) = LIST(K)
12       LIST(K) = NTEMP
13    25 CONTINUE
14       N= N-1
15       IF(N.GT.1) GO TO 20
16    30 WRITE(61,15) (LIST(I), I=1,N)
17       END
```

Answer the following four questions assuming that N = 5.

QUESTION 1   The LIST is sorted to

        (a)   put the even numbers first.

        (b)   put the odd numbers first.

        (c)   put the numbers in increasing order.

        (d)   put the numbers in decreasing order.

        (e)   make no change in the LIST.

This program is a push-down sort. The sort begins by comparing the two values at the top of the list. If a larger value is above a smaller value, they are reversed in the list. As the sort proceeds, larger values are moved downward (and smaller values moved upward) so the largest value encountered moves to the bottom of the list sorted. In the next iteration the list to be sorted is shortened (N= N-1) and the process repeated. This continues until the list has been shortened to one item - at which time the sort is completed. The sorted list will be ordered with the smallest value at the top and the largest value at the bottom. ●

Answer is (c)

QUESTION 2 How many times is the statement in Line 8 executed?

(a) 4
(b) 10
(c) 14
(d) 15
(e) 20

〰〰〰〰〰〰〰〰

The DO loop at line 7 causes line 8 to be executed a number of times for each value of K.
With N=5, K=2, 3, 4, 5.

When   Line 8 is executed

N = 5	4 times
4	3
3	2
2	$\frac{1}{10}$ ●

Answer is (b)

QUESTION 3 The READ statement in Line 5 will cause how many data lines to be read?

(a) 1
(b) 2
(c) 3
(d) 4
(e) 5

〰〰〰〰〰〰〰〰

Answer is (a) ●

QUESTION 4 Which one of the following is untrue?

(a) The same FORMAT statement may be used for both a READ and a WRITE statement.

(b) Line 10 could not be omitted from the program and the same results be obtained.

(c) If the expression in line 15 were changed from (N.GT.1) to (N.GT.2) and line 14 were omitted, the program would not run properly.

(d) An alternate way of writing line 15 is IF(N−1)20,30,20

(e) In this program LIST may contain either INTEGER or REAL numbers.

〰〰〰〰〰〰〰〰

The I-N rules apply to subscripted variables.

Answer is (e) ●

PROGRAMMING 11

Which one of the following lines has a FORTRAN syntax error?

    (a) DIMENSION ABE (4.0)
    (b) REAL K
    (c) PRINT*, 'EIT EXAM'
    (d) DO 20 I = 12,40,3
    (e) IF (X.NE.14.0) NR = NR + 1

An array must be dimensioned as an integer. Thus (a) should read DIMENSION ABE (4).

Answer is (a) ●

PROGRAMMING 12

Select the incorrect FORTRAN statement.

    (a) WRITE(61,10) ((I(I,J),I=1,10),J=1,10)
    (b) WRITE(61,10) (A(I),B(I),I=1,10)
    (c) WRITE(61,10) A,B,C,(D(I),I=1,10)
    (d) WRITE(61,10) (A(I),I=1,10),(B(I),I=5,10)
    (e) WRITE(61,10) I,K,(D(J),J=5,10)

I(I,J) is incorrect. ●

Answer is (a)

PROGRAMMING 13

Select the incorrect FORTRAN statement, if there is one.

    (a) X = FLOAT(I)*Y+Z
    (b) J = INT(Z)+K
    (c) R = FLOAT(K)+FLOAT(L)+M
    (d) A = (A+B)+(C+D)
    (e) All are correct.

Item (d) contains parentheses that do not affect the computation. Thus they are meaningless, but not incorrect. The mixed mode arithmetic in Item (c) is acceptable FORTRAN syntax.

Answer is (e) ●

PROGRAMMING 14

Given: DIMENSION A(3,2,6), B(20)

Which one of the following is correct FORTRAN to EQUIVALENCE A(2,1,4) and B(6)?

        (a)   EQUIVALENCE (A,B)

        (b)   EQUIVALENCE (A(1),B(1))

        (c)   EQUIVALENCE (A(20),B(6))

        (d)   EQUIVALENCE (A(1,1,1),B(1))

        (e)   EQUIVALENCE (A(2,1,4),B(6))

An EQUIVALENCE declaration allows different names to be used for the same storage location in memory. This might be done to reduce the amount of memory required, or for ease in programming. Here the equivalence is to be of a singly subscripted variable and a triply subscripted variable.

The EQUIVALENCE declaration must be written with single subscripted notation irrespective of whether the variables are singly, doubly, or triply subscripted. [The exception would be that item (a) above is an acceptable substitute for item (b).]

Arrays are stored in memory in consecutive sequence. For A(3,2,6) the order of storage is

A(1,1,1), A(2,1,1), A(3,1,1), A(1,2,1), A(2,2,1), A(3,2,1)

A(1,1,2), A(2,1,2), A(3,1,2), A(1,2,2), A(2,2,2), A(3,2,2)

A(1,1,3), A(2,1,3), A(3,1,3), A(1,2,3), A(2,2,3), A(3,2,3)

A(1,1,4), A(2,1,4)   By counting one finds that A(2,1,4) is the 20th storage location. This is to correspond with the 6th storage location for B.

A proper EQUIVALENCE declaration is EQUIVALENCE (A(20),B(6)).
Note that equally correct would be any linkage that aligns the two arrays in the desired manner like   EQUIVALENCE (A(19),B(5))
                     or   EQUIVALENCE (A(15),B(1))

Answer is (c) ●

PROGRAMMING 15

The binary representation   10101   corresponds to which base 10 number?

(a)	3
(b)	16
(c)	21
(d)	42
(e)	10,101

The binary (base 2) number system representation is

$$2^4 \quad 2^3 \quad 2^2 \quad 2^1 \quad 2^0$$

$$16 \quad\ 8 \quad\ 4 \quad\ 2 \quad\ 1$$

binary number $\quad 1 \quad\ \ 0 \quad\ \ 1 \quad\ \ 0 \quad\ \ 1$

$$16 \quad\ 0 \quad\ 4 \quad\ 0 \quad\ 1 = 21\ ●$$

Answer is (c)

PROGRAMMING 16

Consider the following portion of a FORTRAN program

```
      X = 0
      DO 10 I=2,20,4
      DO 10 J=2,20
   10 X = X+1
```

After these statements have been executed, the value of X is

(a)	24
(b)	95
(c)	361
(d)	400
(e)	Not Listed

The outer DO is executed for values of I=2,6,10,14,18 = 5 times.
The inner DO is executed for values of J=2 through 20 = 19 times.

$$19 \times 5 = 95\ ●$$

Answer is (b)

PROGRAMMING 17

In BASIC, which one of the following is untrue?

(a) Every statement normally is placed on a separate line.

(b) Each statement must have a line number.

(c) Blank spaces may be inserted where desired to improve readability of the statement.

(d) Successive statements need not have increasing line numbers.

(e) Each statement must have a BASIC keyword indicating the type of instruction.

Normally successive statements in BASIC *must* have increasing line numbers.

Answer is (d) ●

PROGRAMMING 18

Which one of the following is *not* a BASIC statement?

(a) IF A>B THEN 60

(b) FORMAT (5X,F10.1)

(c) REM SUM

(d) INPUT A,B

(e) LET X=1

Item (b) is a FORTRAN format statement for reading or writing a floating point number. In BASIC, printing of variables is usually done using a PRINT or WRITE statement.

Answer is (b) ●

PROGRAMMING 19

A loop in BASIC begins with a FOR-TO statement. Which statement closes this loop?

(a) NEXT

(b) GOTO

(c) CONTINUE

(d) STOP

(e) END

Answer is (a) ●

PROGRAMMING 20

```
10   READ A,B,C
20   DATA 10,15,20
30   LET D = 30
40   PRINT A+5
50   RUN
```

The error in the above BASIC program is in the line numbered

    (a)   10

    (b)   20

    (c)   30

    (d)   40

    (e)   50

Lines 10 through 40 are simple BASIC statements that are correct. Note that in line 30 LET D = 30 could have been written D = 30 without the LET, since LET is optional in BASIC.

Line 50 is a BASIC command which should not have a line number.

Answer is (e)●

PROGRAMMING 21

Which one of the following lines has a BASIC syntax error?

    (a)   50 DIM YEARS(3,4)

    (b)   60 REM EIT EXAM

    (c)   70 LET = 4

    (d)   80 FOR I=4 TO 25 STEP 2

    (e)   90 A = I + 0.6

LET is properly used as part of an assignment statement. Answer (e), for example, could have been written 90 LET A = I + 0.6. Because it is not needed, LET is probably seldom used. Nevertheless, it remains a Reserved Word in BASIC and cannot be used as a variable name as in (c).

Answer is (c)●

PROGRAMMING 22

Consider the BASIC statement:

$$J = 3 * 4 \wedge 2./5. + 2 * 2$$

When the computer statement above is evaluated, the value of J is nearest to

(a)   9

(b)   13

(c)   14

(d)   20

(e)   32

~~~~~~~~~~~~

The standard order of preference of arithmetic operators is:

Exponentiation

Negation   (assigning a negative sign to an expression)

Multiplication and
Floating Point Division

Integer Division   (represented by the backslash)

Modulo Arithmetic   (represented by MOD)

Addition and
Substraction

Operations within parentheses are performed first. Operations at the same level (Multiplication and Floating Point Division and also Addition and Substraction) are performed in a left to right order.

The statement is evaluated as

$$J = (3*4^2)/5. + 4 = 48./5. + 4$$

$$= 9.6 + 4 = 13.6 \text{ which truncates to } 13 \bullet$$

Answer is (b)

PROGRAMMING 23

If a loop in FORTRAN begins with a IF - THEN statement, which statement closes this loop?

(a)   NEXT
(b)   CONTINUE
(c)   ENDIF
(d)   RETURN
(e)   STOP

~~~~~~~~~~~~

The range of the IF - THEN loop is marked by the ENDIF statement.●

Answer is (c)

PROGRAMMING 24

Consider the following BASIC computer program.

```
10 DIM YEARS (2,3)
20 FOR I = 0 TO 2: FOR J = 0 TO 3
30 YEARS (I,J) = YEARS (I,J) + 1
40 WRITE YEARS (I,J)
50 NEXT J,I
RUN
```

The number of values printed by the program is closest to

(a)   0

(b)   1

(c)   6

(d) 12

(e) 24

In BASIC zero is the first element of an array unless the OPTION BASE statement is used to change it.

Thus YEARS (2,3) has 0, 1, 2 x 0, 1, 2, 3 = 12 elements in it.

The nested FOR-NEXT loops call for all 12 elements to be printed.●

Answer is (d)

PROGRAMMING 25

The largest numerical value printed by the program is closest to

(a)   0

(b)   1

(c)   6

(d) 12

(e) 24

The program prints the value "1" twelve times.●

Answer is (b)

PROGRAMMING 26

The volume of a sphere is

$$V = \frac{4}{3} \pi R^3$$

Which FORTRAN statement statement best represents this situation?

(a)   V = (4/3)πR**3

(b)   V = (4/3)π*R**3

(c)   V = (4/3)*3.14159R**3

(d)   V = (4/3)*3.14159*R**3

(e)   V = 4*3.14159/3.*R**3

In (a) and (b) π is not a FORTRAN character. In (a) through (d) the (4/3) represents integer division which results in a value of 1, and not 1.3333. Answer (e) has a mixed mode multiplication, but FORTRAN 77 permits this with the result of the multiplication being a real number.

Answer is (e)●

PROGRAMMING 27

For the base 10 number of 30, the equivalent binary number is

(a)   1110

(b)   0111

(c)   1111

(d)   11110

(e)   None of the above.

2^5	2^4	2^3	2^2	2^1	2^0	
32	16	8	4	2	1	
	1	1	1	1	0	= 30

Answer is (d).●

13
ENGINEERING ECONOMICS

We begin this chapter with a brief review of engineering economics. The topics cover the full range of what has appeared on past Engineer-In-Training exams, plus four topics that are likely to appear in the future:

Geometric gradient;

Rate of return analysis of three alternatives;

Benefit–cost ratio;

After-tax economic analysis.

There are 26 example problems scattered throughout the engineering economics review. These examples are an integral part of the review and should be worked to completion as you come to them.

CASH FLOW

The field of engineering economics uses mathematical and economic techniques to systematically analyze situations which pose alternative courses of action.

The initial step in engineering economics problems is to resolve a situation, or each alternative course in a given situation, into its favorable and unfavorable consequences or factors. These are then measured in some common unit—usually money. Those factors which cannot readily be reduced to money are called intangible, or irreducible, factors. Intangible or irreducible factors are not included in any monetary analysis but are considered in conjunction with such an analysis when making the final decision on proposed courses of action.

A cash flow table shows the "money consequences" of a situation and its timing. For example, a simple problem might be to list the year-by-year consequences of purchasing and owning a used car:

Year	Cash Flow	
Beginning of first Year 0	−$4500	Car purchased "now" for $4500 cash. The minus sign indicates a disbursement.
End of Year 1	−350	
End of Year 2	−350	Maintenance costs are $350 per year.
End of Year 3	−350	
End of Year 4	−350	
	+2000	The car is sold at the end of the 4th year for $2000. The plus sign represents a receipt of money.

This same cash flow may be represented graphically:

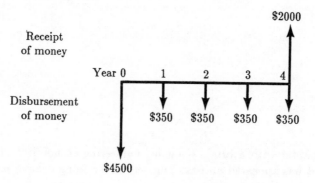

The upward arrow represents a receipt of money, and the downward arrows represent disbursements. The *x*-axis represents the passage of time.

EXAMPLE 1

In January 1990, a firm purchases a used typewriter for $500. Repairs cost nothing in 1990 or 1991. Repairs are $85 in 1992, $130 in 1993, and $140 in 1994. The machine is sold in 1994 for $300. Compute the cash flow table.

Solution

Unless otherwise stated in problems, the customary assumption is a beginning-of-year purchase, followed by end-of-year receipts or disbursements, and an end-of-year resale or salvage value. Thus the typewriter repairs and the typewriter sale are assumed to occur at the end of the year. Letting a minus sign represent a disbursement of money, and a plus sign a receipt of money, we are able to set up this cash flow table:

Year	Cash Flow
Beginning of 1990	−$500
End of 1990	0
End of 1991	0
End of 1992	−85
End of 1993	−130
End of 1994	+160

Notice that at the end of 1994, the cash flow table shows +160 which is the net sum of −140 and +300. If we define Year 0 as the beginning of 1990, the cash flow table becomes:

Year	Cash Flow
0	−$500
1	0
2	0
3	−85
4	−130
5	+160

From this cash flow table, the definitions of Year 0 and Year 1 become clear. Year 0 is defined as the *beginning* of Year 1. Year 1 is the *end* of Year 1. Year 2 is the *end* of Year 2, and so forth.

TIME VALUE OF MONEY

When the money consequences of an alternative occur in a short period of time—say, less than one year—we might simply add up the various sums of money and obtain the net result. But we cannot treat money this same way over longer periods of time. This is because money today is not the same as money at some future time.

Consider this question: Which would you prefer, $100 today or the assurance of receiving $100 a year from now? Clearly, you would prefer the $100 today. If you had the money today, rather than a year from now, you could use it for the year. And if you had no use for it, you could lend it to someone who would pay interest for the privilege of using your money for the year.

EQUIVALENCE

In the preceding section we saw that money at different points in time (for example, $100 today or $100 one year hence) may be equal in the sense that they both are $100, but $100 a year hence is *not* an acceptable substitute for $100 today. When we have acceptable substitutes, we say they are *equivalent* to each other. Thus at 8% interest, $108 a year hence is equivalent to $100 today.

EXAMPLE 2
At a 10% per year interest rate, $500 now is *equivalent* to how much three years hence?

Solution
$500 now will increase by 10% in each of the three years.

$$
\begin{aligned}
\text{Now} &= & &\$500 \\
\text{End of 1st year} &= 500 + 10\%(500) &=& 550 \\
\text{End of 2nd year} &= 550 + 10\%(550) &=& 605 \\
\text{End of 3rd year} &= 605 + 10\%(605) &=& 665.50
\end{aligned}
$$

Thus $500 now is *equivalent* to $665.50 at the end of three years.

Equivalence is an essential factor in engineering economic analysis. Suppose we wish to select the better of two alternatives. First, we must compute their cash flows. An example would be:

| | Alternative | |
Year	A	B
0	−$2000	−$2800
1	+800	+1100
2	+800	+1100
3	+800	+1100

The larger investment in Alternative B results in larger subsequent benefits, but we have no direct way of knowing if Alternative B is better than Alternative A. Therefore we do not know which alternative should be selected. To make a decision we must resolve the alternatives into *equivalent* sums so they may be compared accurately and a decision made.

COMPOUND INTEREST FACTORS

To facilitate equivalence computations a series of compound interest factors will be derived and their use illustrated.

Symbols

i = Interest rate per interest period. In equations the interest rate is stated as a decimal (that is, 8% interest is 0.08).

n = Number of interest periods.

P = A present sum of money.

F = A future sum of money. The future sum F is an amount, n interest periods from the present, that is equivalent to P with interest rate i.

A = An end-of-period cash receipt or disbursement in a uniform series continuing for n periods, the entire series equivalent to P or F at interest rate i.

G = Uniform period-by-period increase in cash flows; the arithmetic gradient.

g = Uniform *rate* of period-by-period increase in cash flows; the geometric gradient.

Functional Notation

	To Find	Given	Functional Notation
Single Payment			
Compound Amount Factor	F	P	$(F/P,i,n)$
Present Worth Factor	P	F	$(P/F,i,n)$

FUNCTIONAL NOTATION, continued

	To Find	Given	Functional Notation
Uniform Payment Series			
Sinking Fund Factor	A	F	$(A/F,i,n)$
Capital Recovery Factor	A	P	$(A/P,i,n)$
Compound Amount Factor	F	A	$(F/A,i,n)$
Present Worth Factor	P	A	$(P/A,i,n)$
Arithmetic Gradient			
Gradient Uniform Series	A	G	$(A/G,i,n)$
Gradient Present Worth	P	G	$(P/G,i,n)$

From the table above we can see that the functional notation scheme is based on writing (To Find / Given, i,n). Thus, if we wished to find the future sum F, given a uniform series of receipts A, the proper compound interest factor to use would be $(F/A,i,n)$.

Single Payment Formulas

Suppose a present sum of money P is invested for one year at interest rate i. At the end of the year, we receive back our initial investment P together with interest equal to Pi or a total amount $P + Pi$. Factoring P, the sum at the end of one year is $P(1 + i)$. If we agree to let our investment remain for subsequent years, the progression is as follows:

	Amount at Beginning of Period	+	Interest for the Period	=	Amount at End of the Period
1st year	P	$+$	Pi	$=$	$P(1 + i)$
2nd year	$P(1 + i)$	$+$	$Pi(1 + i)$	$=$	$P(1 + i)^2$
3rd year	$P(1 + i)^2$	$+$	$Pi(1 + i)^2$	$=$	$P(1 + i)^3$
nth year	$P(1 + i)^{n-1}$	$+$	$Pi(1 + i)^{n-1}$	$=$	$P(1 + i)^n$

The present sum P increases in n periods to $P(1 + i)^n$. This gives us a relationship between a present sum P and its equivalent future sum F:

Future Sum = (Present Sum)$(1 + i)^n$
$$F = P(1 + i)^n$$

This is the Single Payment Compound Amount formula. In functional notation it is written:

$$F = P(F/P,i,n)$$

The relationship may be rewritten as:

Present Sum = (Future Sum)$(1 + i)^{-n}$
$$P = F(1 + i)^{-n}$$

This is the Single Payment Present Worth formula. It is written:

$$P = F(P/F,i,n)$$

EXAMPLE 3

At a 10% per year interest rate, $500 now is *equivalent* to how much three years hence?

Solution
This problem was solved in Example 2. Now it can be solved using a single payment formula.

$P = 500
$n = 3$ years
$i = 10\%$
$F = $ unknown

$$F = P(1 + i)^n = 500(1 + 0.10)^3 = \$665.50$$

This problem may also be solved using the Compound Interest Tables.

$$F = P(F/P,i,n) = 500(F/P,10\%,3)$$

From the 10% Compound Interest Table, read $(F/P,10\%,3) = 1.331$.

$$F = 500(F/P,10\%,3) = 500(1.331) = \$665.50$$

EXAMPLE 4

To raise money for a new business, a man asks you to loan him some money. He offers to pay you $3000 at the end of four years. How much should you give him now if you want 12% interest per year on your money?

Solution

$P = $ unknown
$n = 4$ years
$i = 12\%$
$F = 3000

$$P = F(1 + i)^{-n} = 3000(1 + 0.12)^{-4} = \$1906.55$$

Alternate computation using Compound Interest Tables:

$$P = F(P/F,i,n) = 3000(P/F,12\%,4)$$
$$= 3000(0.6355) = \$1906.50$$

Note that the solution based on the Compound Interest Table is slightly different from the exact solution using a hand calculator. In economic analysis, the Compound Interest Tables are always considered to be sufficiently accurate.

Uniform Payment Series Formulas

Consider the following situation:

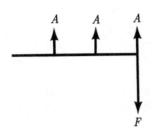

$A = $ End-of-period cash receipt or disbursement in a uniform series continuing for n periods.

$F = $ A future sum of money.

Using the single payment compound amount factor, we can write an equation for F in terms of A:

$$F = A + A(1 + i) + A(1 + i)^2 \qquad (1)$$

In our situation, with $n = 3$, Equation (1) may be written in a more general form:

$$F = A + A(1 + i) + A(1 + i)^{n-1} \qquad (2)$$

Multiply Eq. (2) by $(1 + i)$: $\quad (1 + i)F = A(1 + i) + A(1 + i)^{n-1} + A(1 + i)^n \qquad (3)$

Write Eq. (2): $\qquad\qquad\qquad - \quad F = A + A(1 + i) + A(1 + i)^{n-1} \qquad (2)$

$(3) - (2)$: $\qquad\qquad\qquad\qquad iF = -A + A(1 + i)^n$

$$F = A\left(\frac{(1 + i)^n - 1}{i}\right)$$

Uniform Series Compound Amount formula

Solving this equation for A:

$$A = F\left(\frac{i}{(1 + i)^n - 1}\right) \qquad \text{Uniform Series Sinking Fund formula}$$

Since $F = P(1 + i)^n$, we can substitute this expression for F in the equation and obtain:

$$A = P\left(\frac{i(1 + i)^n}{(1 + i)^n - 1}\right) \qquad \text{Uniform Series Capital Recovery formula}$$

Solving the equation for P:

$$P = A\left(\frac{(1 + i)^n - 1}{i(1 + i)^n}\right) \qquad \text{Uniform Series Present Worth formula}$$

In functional notation, the uniform series factors are:

Compound Amount	$(F/A,i,n)$
Sinking Fund	$(A/F,i,n)$
Capital Recovery	$(A/P,i,n)$
Present Worth	$(P/A,i,n)$

EXAMPLE 5

If $100 is deposited at the end of each year in a savings account that pays 6% interest per year, how much will be in the account at the end of five years?

Solution

$$A = \$100$$
$$F = \text{unknown}$$
$$n = 5 \text{ years}$$
$$i = 6\%$$

$$F = A(F/A,i,n) = 100(F/A,6\%,5)$$
$$= 100(5.637) = \$563.70$$

EXAMPLE 6

A woman wishes to make a uniform deposit every three months to her savings account so that at the end of 10 years she will have $10,000 in the account. If the account earns 6% annual interest, compounded quarterly, how much should she deposit each three months?

Solution

$$F = \$10,000$$
$$A = \text{unknown}$$
$$n = 40 \text{ quarterly deposits}$$
$$i = 1\tfrac{1}{2}\% \text{ per quarter year}$$

Note that i, the interest rate per interest period, is $1\tfrac{1}{2}\%$, and there are 40 deposits.

$$A = F(A/F,i,n) = 10,000(A/F,1\tfrac{1}{2}\%,40)$$
$$= 10,000(0.0184) = \$184$$

EXAMPLE 7

An individual is considering the purchase of a used automobile. The total price is $6200 with $1240 as a downpayment and the balance paid in 48 equal monthly payments with interest at 1% per month. The payments are due at the end of each month. Compute the monthly payment.

Solution

The amount to be repaid by the 48 monthly payments is the cost of the automobile *minus* the $1240 downpayment.

$P = $ \$4960
$A = $ unknown
$n = $ 48 monthly payments
$i = $ 1% per month

$$A = P(A/P,i,n) = 4960(A/P,1\%,48)$$
$$= 4960(0.0263) = \$130.45$$

EXAMPLE 8
A couple sold their home. In addition to cash, they took a mortgage on the house. The mortgage will be paid off by monthly payments of \$232.50 for 10 years. The couple decides to sell the mortgage to a local bank. The bank will buy the mortgage, but requires a 1% per month interest rate on their investment. How much will the bank pay for the mortgage?

Solution

$A = $ \$232.50
$n = $ 120 months
$i = $ 1% per month
$P = $ unknown

$$P = A(P/A,i,n) = 232.50(P/A,1\%,120)$$
$$= 232.50(69.701) = \$16,205.48$$

Arithmetic Gradient

At times one will encounter a situation where the cash flow series is not a constant amount A. Instead it is an increasing series like:

This cash flow may be resolved into two components:

We can compute the value of P^* as equal to P' plus P. We already have an equation for P':

$$P' = A(P/A,i,n)$$

The value for P in the right-hand diagram is:

$$P = G\left(\frac{(1 + i)^n - in - 1}{i^2(1 + i)^n}\right)$$

This is the Arithmetic Gradient Present Worth formula. In functional notation, the relationship is $P = G(P/G,i,n)$.

EXAMPLE 9

The maintenance on a machine is expected to be \$155 at the end of the first year, and increasing \$35 each year for the following seven years. What present sum of money would need to be set aside now to pay the maintenance for the eight-year period? Assume 6% interest.

Solution

$$P = 155(P/A,6\%,8) + 35(P/G,6\%,8)$$
$$= 155(6.210) + 35(19.841) = \$1656.99$$

In the gradient series, if instead of the present sum P, an equivalent uniform series A is desired, the problem becomes:

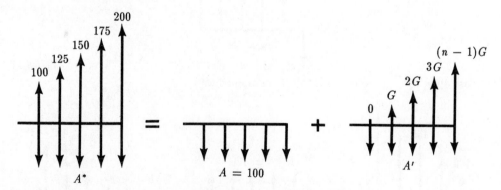

The relationship between A' and G in the right-hand diagram is:

$$A' = G\left(\frac{(1 + i)^n - in - 1}{i(1 + i)^n - i}\right)$$

In functional notation, the Arithmetic Gradient (to) Uniform Series factor is:

$$A = G(A/G, i, n)$$

It is important to note carefully the diagrams for the two arithmetic gradient series factors. In both cases the first term in the arithmetic gradient series is zero and the last term is $(n - 1)G$. But we use n in the equations and functional notation. The derivations (not shown here) were done on this basis and the arithmetic gradient series Compound Interest Tables are computed this way.

EXAMPLE 10

For the situation in Example 9, we wish now to know the uniform annual maintenance cost. Compute an equivalent A for the maintenance costs.

Solution

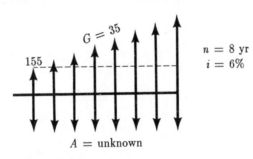

$$n = 8 \text{ yr}$$
$$i = 6\%$$

$$A = \text{unknown}$$

Equivalent uniform annual maintenance cost:

$$
\begin{aligned}
A &= 155 + 35(A/G, 6\%, 8) \\
  &= 155 + 35(3.195) = \$266.83
\end{aligned}
$$

Geometric Gradient

The arithmetic gradient is applicable where the period-by-period change in the cash flow is a uniform amount. There are other situations where the period-by-period change is a *uniform rate*, g. A diagram of this situation is:

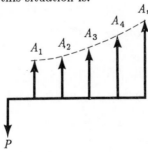

where $A_n = A_1(1 + g)^{n-1}$

$g = $ Uniform rate of period-by-period change; the geometric gradient stated as a decimal ($8\% = 0.08$).

$A_1 = $ Value of A at Year 1.

$A_n = $ Value of A at any Year n.

Geometric Series Present Worth Formulas:

When $i = g$, $P = A_1\left(n(1 + i)^{-1}\right)$

When $i \neq g$, $P = A_1\left(\dfrac{1 - (1 + g)^n(1 + i)^{-n}}{i - g}\right)$

EXAMPLE 11

It is likely that airplane tickets will increase 8% in each of the next four years. The cost of a plane ticket at the end of the first year will be $180. How much money would need to be placed in a savings account now to have money to pay a student's travel home at the end of each year for the next four years? Assume the savings account pays 5% annual interest.

Solution

The problem describes a geometric gradient where $g = 8\%$ and $i = 5\%$.

$$P = A_1\left(\frac{1 - (1 + g)^n(1 + i)^{-n}}{i - g}\right)$$

$$= 180.00\left(\frac{1 - (1.08)^4(1.05)^{-4}}{0.05 - 0.08}\right) = 180.00\left(\frac{-0.119278}{-0.03}\right) = \$715.67$$

Thus, $715.67 would need to be deposited now.

As a check, the problem can be solved without using the geometric gradient:

Year		Ticket
1	$A_1 =$	$180.00
2	$A_2 = 180.00 + 8\%(180.00) =$	194.40
3	$A_3 = 194.40 + 8\%(194.40) =$	209.95
4	$A_4 = 209.95 + 8\%(209.95) =$	226.75

$P = 180.00(P/F,5\%,1) + 194.40(P/F,5\%,2) + 209.95(P/F,5\%,3) + 226.75(P/F,5\%,4)$

$= 180.00(0.9524) + 194.40(0.9070) + 209.95(0.8638) + 226.75(0.8227)$

$= \$715.66$

NOMINAL AND EFFECTIVE INTEREST

Nominal interest is the annual interest rate without considering the effect of any compounding.

Effective interest is the annual interest rate taking into account the effect of any compounding during the year.

Frequently an interest rate is described as an annual rate, even though the interest period may be something other than one year. A bank may pay 1–1/2% interest on the amount in a savings account every three months. The *nominal* interest rate in this situation is 6% (4 × 1–1/2% = 6%). But if you deposited $1000 in such an account, would you have 106%(1000) = $1060 in the account at the end of one year? The answer is no, you would have more. The amount in the account would increase as follows:

Amount in Account

At beginning of year = $1000.00

End of 3 months: $1000.00 + 1\frac{1}{2}\%(1000.00) = 1015.00$

End of 6 months: $1015.00 + 1\frac{1}{2}\%(1015.00) = 1030.23$

End of 9 months: $1030.23 + 1\frac{1}{2}\%(1030.23) = 1045.68$

End of one year: $1045.68 + 1\frac{1}{2}\%(1045.68) = 1061.37$

The actual interest rate on the $1000 would be the interest, $61.37, divided by the original $1000, or 6.137%. We call this the *effective* interest rate.

Effective interest rate = $(1 + i)^m - 1$, where

i = Interest rate per interest period;
m = Number of compoundings per year.

EXAMPLE 12

A bank charges $1\frac{1}{2}\%$ per month on the unpaid balance for purchases made on its credit card. What nominal interest rate is it charging? What effective interest rate?

Solution
The nominal interest rate is simply the annual interest ignoring compounding, or $12(1\frac{1}{2}\%) = 18\%$.

Effective interest rate = $(1 + 0.015)^{12} - 1 = 0.1956 = 19.56\%$

SOLVING ECONOMIC ANALYSIS PROBLEMS

The techniques presented so far illustrate how to convert single amounts of money, and uniform or gradient series of money, into some equivalent sum at another point in time. These compound interest computations are an essential part of economic analysis problems.

The typical situation is that we have a number of alternatives and the question is, which alternative should be selected? The customary method of solution is to resolve each of the alternatives into some common form and then choose the best alternative (taking both the monetary and intangible factors into account).

Criteria

Economic analysis problems inevitably fall into one of three categories:

1. Fixed Input   The amount of money or other input resources is fixed.

   Example: A project engineer has a budget of $450,000 to overhaul a plant.

2. Fixed Output   There is a fixed task, or other output to be accomplished.

   Example: A mechanical contractor has been awarded a fixed price contract to air-condition a building.

3. Neither Input   This is the general situation where neither the amount
   nor Output Fixed   of money or other inputs, nor the amount of benefits
                       or other outputs are fixed.

   Example: A consulting engineering firm has more work available than it can handle. It is considering paying the staff for working evenings to increase the amount of design work it can perform.

There are five major methods of comparing alternatives: present worth; future worth; annual cost; rate of return; and benefit–cost ratio. These are presented in the following sections.

PRESENT WORTH

In present worth analysis, the approach is to resolve all the money consequences of an alternative into an equivalent present sum. For the three categories given above, the criteria are:

Category	*Present Worth Criterion*
Fixed Input	Maximize the Present Worth of benefits or other outputs.
Fixed Output	Minimize the Present Worth of costs or other inputs.
Neither Input nor Output Fixed	Maximize [Present Worth of benefits *minus* Present Worth of costs] or, stated another way: Maximize Net Present Worth.

Application of Present Worth

Present worth analysis is most frequently used to determine the present value of future money receipts and disbursements. We might want to know, for example, the present worth of an income producing property, like an oil well. This should provide an estimate of the price at which the property could be bought or sold.

An important restriction in the use of present worth calculations is that there must be a common analysis period when comparing alternatives. It would be incorrect, for example, to compare the present worth (PW) of cost of Pump *A*, expected to last 6 years, with the PW of cost of Pump *B*, expected to last 12 years.

Improper Present Worth Comparison

In situations like this, the solution is either to use some other analysis technique* or to restructure the problem so there is a common analysis period. In the example above, a customary assumption would be that a pump is needed for 12 years and that Pump *A* will be replaced by an identical Pump *A* at the end of 6 years. This gives a 12-year common analysis period.

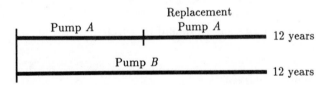

Correct Present Worth Comparison

This approach is easy to use when the different lives of the alternatives have a practical least common multiple life. When this is not true (for example, life of *J* equals 7 years and the life of *K* equals 11 years), some assumptions must be made to select a suitable common analysis period, or the present worth method should not be used.

EXAMPLE 13
Machine *X* has an initial cost of $10,000, annual maintenance of $500 per year, and no salvage value at the end of its four-year useful life. Machine *Y* costs $20,000. The first year there is no maintenance cost. The second year, maintenance is $100, and increases $100 per year in subsequent years. The machine has an anticipated $5000 salvage value at the end of its 12-year useful life.

  If interest is 8%, which machine should be selected?

Solution
The analysis period is not stated in the problem. Therefore we select the least common multiple of the lives, or 12 years, as the analysis period.

---

*Generally the annual cost method is suitable in these situations.

Present Worth of Cost of 12 years of Machine X

$$= 10{,}000 + 10{,}000(P/F{,}8\%{,}4) + 10{,}000(P/F{,}8\%{,}8) + 500(P/A{,}8\%{,}12)$$
$$= 10{,}000 + 10{,}000(0.7350) + 10{,}000(0.5403) + 500(7.536)$$
$$= \$26{,}521$$

Present Worth of Cost of 12 years of Machine Y

$$= 20{,}000 + 100(P/G{,}8\%{,}12) - 5000(P/F{,}8\%{,}12)$$
$$= 20{,}000 + 100(34.634) - 5000(0.3971)$$
$$= \$21{,}478$$

Choose Machine Y with its smaller PW of Cost.

EXAMPLE 14
Two alternatives have the following cash flows:

	Alternative	
Year	A	B
0	−$2000	−$2800
1	+800	+1100
2	+800	+1100
3	+800	+1100

At a 5% interest rate, which alternative should be selected?

Solution
Solving by Present Worth analysis:

Net Present Worth (NPW) = PW of benefits − PW of cost

$$NPW_A = 800(P/A{,}5\%{,}3) - 2000$$
$$= 800(2.723) - 2000$$
$$= +178.40$$

$$NPW_B = 1100(P/A{,}5\%{,}3) - 2800$$
$$= 1100(2.723) - 2800$$
$$= +195.30$$

To maximize NPW, choose Alternative B.

Capitalized Cost

In the special situation where the analysis period is infinite ($n = \infty$), an analysis of the present worth of cost is called *capitalized cost*. There are a few public projects where the analysis period is infinity. Other examples would be permanent endowments and cemetery perpetual care.

When n equals infinity, a present sum P will accrue interest of Pi for every future interest period. For the principal sum P to continue undiminished (an essential requirement for *n* equal to infinity), the end-of-period sum *A* that can be disbursed is *Pi*.

$$P \longrightarrow P + Pi \longrightarrow P + Pi \longrightarrow P + Pi \longrightarrow \cdots$$
$$\qquad\qquad A \qquad\qquad A \qquad\qquad A$$

When $n = \infty$, the fundamental relationship between *P*, *A*, and *i* is:

$$A = Pi$$

Some form of this equation is used whenever there is a problem with an infinite analysis period.

EXAMPLE 15
In his will, a man wishes to establish a perpetual trust to provide for the maintenance of a small local park. If the annual maintenance is $7500 per year and the trust account can earn 5% interest, how much money must be set aside in the trust?

Solution
When $n = \infty$, $A = Pi$ or $P = \dfrac{A}{i}$

Capitalized cost $P = \dfrac{A}{i} = \dfrac{\$7500}{0.05} = \$150,000$

FUTURE WORTH

In present worth analysis, the comparison is made in terms of the equivalent present costs and benefits. But the analysis need not be made at the present time, it could be made at any point in time: past, present, or future. Although the numerical calculations may look different, the decision is unaffected by the point in time selected. Of course, there are situations where we do want to know what the future situation will be if we take some particular course of action now. When an analysis is made based on some future point in time, it is called future worth analysis.

Category	*Future Worth Criterion*
Fixed Input	Maximize the Future Worth of benefits or other outputs.
Fixed Output	Minimize the Future Worth of costs or other inputs.
Neither Input nor Output Fixed	Maximize [Future Worth of benefits *minus* Future Worth of costs] or, stated another way: Maximize Net Future Worth.

EXAMPLE 16
Two alternatives have the following cash flows:

	Alternative	
Year	A	B
0	−$2000	−$2800
1	+800	+1100
2	+800	+1100
3	+800	+1100

At a 5% interest rate, which alternative should be selected?

Solution
In Example 14, this problem was solved by Present Worth analysis at Year 0. Here it will be solved by Future Worth analysis at the end of Year 3.

Net Future Worth (NFW) = FW of benefits − FW of cost

$$
\begin{aligned}
\text{NFW}_A &= 800(F/A,5\%,3) - 2000(F/P,5\%,3) \\
&= 800(3.152) - 2000(1.158) \\
&= +205.60
\end{aligned}
$$

$$
\begin{aligned}
\text{NFW}_B &= 1100(F/A,5\%,3) - 2800(F/P,5\%,3) \\
&= 1100(3.152) - 2800(1.158) \\
&= +224.80
\end{aligned}
$$

To maximize NFW, choose Alternative *B*.

ANNUAL COST

The annual cost method is more accurately described as the method of Equivalent Uniform Annual Cost (EUAC) or, where the computation is of benefits, the method of Equivalent Uniform Annual Benefits (EUAB).

Criteria

For each of the three possible categories of problems, there is an annual cost criterion for economic efficiency.

Category	*Annual Cost Criterion*
Fixed Input	Maximize the Equivalent Uniform Annual Benefits. That is, maximize EUAB.
Fixed Output	Minimize the Equivalent Uniform Annual Cost. That is, minimize EUAC.
Neither Input nor Output Fixed	Maximize [EUAB − EUAC].

Application of Annual Cost Analysis

In the section on present worth, we pointed out that the present worth method requires that there be a common analysis period for all alternatives. This same restriction does not apply in all annual cost calculations, but it is important to understand the circumstances that justify comparing alternatives with different service lives.

Frequently an analysis is to provide for a more or less continuing requirement. One might need to pump water from a well, for example, as a continuing requirement. Regardless of whether the pump has a useful service life of 6 years or 12 years, we would select the one whose annual cost is a minimum. And this would still be the case if the pump useful lives were the more troublesome 7 and 11 years, respectively. Thus, if we can assume a continuing need for an item, an annual cost comparison among alternatives of differing service lives is valid.

The underlying assumption made in these situations is that when the shorter-lived alternative has reached the end of its useful life, it can be replaced with an identical item with identical costs, and so forth. This means the EUAC of the initial alternative is equal to the EUAC for the continuing series of replacements.

If, on the other hand, there is a specific requirement in some situation to pump water for 10 years, then each pump must be evaluated to see what costs will be incurred during the analysis period and what salvage value, if any, may be recovered at the end of the analysis period. The annual cost comparison needs to consider the actual circumstances of the situation.

Examination problems are often readily solved by the annual cost method. And the underlying "continuing requirement" is often present, so that an annual cost comparison of unequal-lived alternatives is an appropriate method of analysis.

EXAMPLE 17

Consider the following alternatives:

	A	B
First cost	$5000	$10,000
Annual maintenance	500	200
End-of-useful-life salvage value	600	1000
Useful life	5 years	15 years

Based on an 8% interest rate, which alternative should be selected?

Solution
Assuming both alternatives perform the same task and there is a continuing requirement, the goal is to minimize EUAC.

Alternative A:

$$EUAC = 5000(A/P,8\%,5) + 500 - 600(A/F,8\%,5)$$
$$= 5000(0.2505) + 500 - 600(0.1705) = \$1650$$

Alternative B:

$$EUAC = 10,000(A/P,8\%,15) + 200 - 1000(A/F,8\%,15)$$

$$= 10,000(0.1168) + 200 - 1000(0.0368) = \$1331$$

To minimize EUAC, select Alternative *B*.

RATE OF RETURN

A typical situation is a cash flow representing the costs and benefits. The rate of return may be defined as the interest rate where

PW of cost = PW of benefits,

EUAC = EUAB,

or PW of cost − PW of benefits = 0.

EXAMPLE 18

Compute the rate of return for the investment represented by the following cash flow:

Year	Cash Flow
0	−$595
1	+250
2	+200
3	+150
4	+100
5	+50·

Solution

This declining arithmetic gradient series may be separated into two cash flows for which compound interest factors are available:

Note that the gradient series factors are based on an *increasing* gradient. Here, the declining cash flow is solved by subtracting an increasing arithmetic gradient, as indicated by the diagram.

PW of cost − PW of benefits = 0

$$595 - [250(P/A,i,5) - 50(P/G,i,5)] = 0$$

Try $i = 10\%$:

$$595 - [250(3.791) - 50(6.862)] = -9.65$$

Try $i = 12\%$:

$$595 - [250(3.605) - 50(6.397)] = +13.60$$

The rate of return is between 10% and 12%. It may be computed more accurately by linear interpolation:

$$\text{Rate of return} = 10\% + (2\%)\left(\frac{9.65 - 0}{13.60 + 9.65}\right) = 10.83\%$$

Rate of Return Criterion for Two Alternatives

Compute the incremental rate of return on the cash flow representing the difference between the two alternatives. Since we want to look at increments of *investment*, the cash flow for the difference between the alternatives is computed by taking the higher initial-cost alternative *minus* the lower initial-cost alternative. If the incremental rate of return is greater than or equal to the predetermined minimum attractive rate of return (MARR), choose the higher-cost alternative; otherwise, choose the lower-cost alternative.

EXAMPLE 19

Two alternatives have the following cash flows:

	Alternative	
Year	A	B
0	−$2000	−$2800
1	+800	+1100
2	+800	+1100
3	+800	+1100

If 5% is considered the minimum attractive rate of return (MARR), which alternative should be selected?

Solution

These two alternatives were previously examined in Examples 14 and 16 by present worth and future worth analysis. This time, the alternatives will be resolved using rate of return analysis.

Note that the problem statement specifies a 5% minimum attractive rate of raturn (MARR), while Examples 14 and 16 referred to a 5% interest rate. These are really two different ways of saying the same thing: the minimum acceptable time value of money is 5%.

First, tabulate the cash flow that represents the increment of investment between the alternatives. This is done by taking the higher initial-cost alternative minus the lower initial-cost alternative:

		ative	Difference between Alternatives
Year	A	B	B − A
0	−$2000	−$2800	−$800
1	+800	+1100	+300
2	+800	+1100	+300
3	+800	+1100	+300

Then compute the rate of return on the increment of investment represented by the difference between the alternatives:

$$\text{PW of cost} = \text{PW of benefits}$$
$$800 = 300(P/A,i,3)$$
$$(P/A,i,3) = \frac{800}{300} = 2.67$$
$$i \approx 6.1\%$$

Since the incremental rate of return exceeds the 5% MARR, the increment of investment is desirable. Choose the higher-cost Alternative B.

Before leaving this example problem, one should note something that relates to the rates of return on Alternative A and on Alternative B. These rates of return, if computed, are:

	Rate of Return
Alternative A	9.7%
Alternative B	8.7%

The correct answer to this problem has been shown to be Alternative B, and this is true even though Alternative A has a higher rate of return. The higher-cost alternative may be thought of as the lower-cost alternative, plus the increment of investment between them. Looked at this way, the higher-cost Alternative B is equal to the desirable lower-cost Alternative A plus the desirable differences between the alternatives.

The important conclusion is that computing the rate of return for each alternative does not provide the basis for choosing between alternatives. Instead, incremental analysis is required.

EXAMPLE 20

Consider the following:

	Alternative	
Year	A	B
0	−$200.0	−$131.0
1	+77.6	+48.1
2	+77.6	+48.1
3	+77.6	+48.1

If the minimum attractive rate of return (MARR) is 10%, which alternative should be selected?

Solution

To examine the increment of investment between the alternatives, we will examine the higher initial-cost alternative minus the lower initial-cost alternative, or $A - B$.

	Alternative		Increment
Year	A	B	$A - B$
0	−$200.0	−$131.0	−$69.0
1	+77.6	+48.1	+29.5
2	+77.6	+48.1	+29.5
3	+77.6	+48.1	+29.5

Solve for the incremental rate of return:

$$\text{PW of cost} = \text{PW of benefits}$$
$$69.0 = 29.5(P/A,i,3)$$
$$(P/A,i,3) = \frac{69.0}{29.5} = 2.339$$

From Compound Interest Tables, the incremental rate of return is between 12% and 15%. This is a desirable increment of investment hence we select the higher initial-cost Alternative *A*.

Rate of Return Criterion for Three or More Alternatives

When there are three or more mutually exclusive alternatives, one must proceed following the same general logic presented for two alternatives. The components of incremental analysis are:

1. Compute the rate of return for each alternative. Reject any alternative where the rate of return is less than the given MARR. (This step is not essential, but helps to immediately identify unacceptable alternatives.)

2. Rank the remaining alternatives in their order of increasing initial cost.

3. Examine the increment of investment between the two lowest-cost alternatives as described for the two-alternative problem. Select the best of the two alternatives and reject the other one.

4. Take the preferred alternative from Step 3. Consider the next higher initial-cost alternative and proceed with another two-alternative comparison.

5. Continue until all alternatives have been examined and the best of the multiple alternatives has been identified.

EXAMPLE 21
Consider the following:

	Alternative	
Year	A	B
0	−$200.0	−$131.0
1	+77.6	+48.1
2	+77.6	+48.1
3	+77.6	+48.1

If the minimum attractive rate of return (MARR) is 10%, which alternative, if any, should be selected?

Solution
One should carefully note that this is a *three*-alternative problem where the alternatives are A, B, and "Do Nothing."
   In this solution we will skip Step 1. Reorganize the problem by placing the alternatives in order of increasing initial cost:

	Do	Alternative	
Year	Nothing	B	A
0	0	−$131.0	−$200.0
1	0	+48.1	+77.6
2	0	+48.1	+77.6
3	0	+48.1	+77.6

Examine the "B − Do Nothing" increment of investment:

Year	B − Do Nothing	
0	−$131.0 − 0 =	−$131.0
1	+48.1 − 0 =	+48.1
2	+48.1 − 0 =	+48.1
3	+48.1 − 0 =	+48.1

Solve for the incremental rate of return:

$$PW \text{ of cost} = PW \text{ of benefits}$$
$$131.0 = 48.1(P/A,i,3)$$
$$(P/A,i,3) = \frac{131.0}{48.1} = 2.723$$

From Compound Interest Tables, the incremental rate of return = 5%. Since the incremental rate of return is less than 10%, the B − Do Nothing increment is not desirable. Reject Alternative B.
   Next, consider the increment of investment between the two remaining alternatives:

Year	A − Do Nothing	
0	−$200.0 − 0 =	−$200.0
1	+77.6 − 0 =	+77.6
2	+77.6 − 0 =	+77.6
3	+77.6 − 0 =	+77.6

Solve for the incremental rate of return:

PW of cost = PW of benefits
$$200.0 = 77.6(P/A,i,3)$$
$$(P/A,i,3) = \frac{200.0}{77.6} = 2.577$$

The incremental rate of return is 8%. Since the rate of return on the A — Do Nothing increment of investment is less than the desired 10%, reject the increment by rejecting Alternative A. We select the remaining alternative: Do nothing!

If you have not already done so, you should go back to Example 20 and see how the slightly changed wording of the problem radically altered it. Example 20 required the choice between two undesirable alternatives. Example 21 adds the Do-nothing alternative which is superior to A or B.

EXAMPLE 22

Consider four mutually exclusive alternatives:

	Alternative			
	A	*B*	*C*	*D*
Initial Cost	$400.0	$100.0	$200.0	$500.0
Uniform Annual Benefit	100.9	27.7	46.2	125.2

Each alternative has a five-year useful life and no salvage value. If the minimum attractive rate of return (MARR) is 6%, which alternative should be selected?

Solution

Mutually exclusive is where selecting one alternative precludes selecting any of the other alternatives. This is the typical "textbook" situation. The solution will follow the several steps in incremental analysis.

1. The rate of return is computed for the four alternatives.

Alternative	*A*	*B*	*C*	*D*
Computed rate of return	8.3%	11.9%	5%	8%

Since Alternative C has a rate of return less than the MARR, it may be eliminated from further consideration.

2. Rank the remaining alternatives in order of increasing intial cost and examine the increment between the two lowest cost alternatives.

Alternative	*B*	*A*	*D*
Initial Cost	$100.0	$400.0	$500.0
Uniform Annual Benefit	27.7	100.9	125.2

	A − *B*
Δ Initial Cost	$300.0
Δ Uniform Annual Benefit	73.2
Computed Δ rate of return	7%

Since the incremental rate of return exceeds the 6% MARR, the increment of investment is desirable. Alternative *A* is the better alternative.

3. Take the preferred alternative from the previous step and consider the next higher-cost alternative. Do another two-alternative comparison.

	D − A
Δ Initial Cost	$100.0
Δ Uniform Annual Benefit	24.3
Computed Δ rate of return	6.9%

The incremental rate of return exceeds MARR, hence the increment is desirable. Alternative *D* is preferred over Alternative *A*.

Conclusion: Select Alternative *D*. Note that once again the alternative with the highest rate of return (Alt. *B*) is *not* the proper choice.

BENEFIT-COST RATIO

Generally in public works and governmental economic analyses, the dominant method of analysis is called benefit–cost ratio. It is simply the ratio of benefits divided by costs, taking into account the time value of money.

$$\text{B/C} = \frac{\text{PW of benefits}}{\text{PW of cost}} = \frac{\text{Equivalent Uniform Annual Benefits}}{\text{Equivalent Uniform Annual Cost}}$$

For a given interest rate, a B/C ratio ≥ 1 reflects an acceptable project. The method of analysis using B/C ratio is parallel to that of rate of return analysis. The same kind of incremental analysis is required.

B/C Ratio Criterion for Two Alternatives

Compute the incremental B/C ratio for the cash flow representing the increment of investment between the higher initial-cost alternative and the lower initial-cost alternative. If this incremental B/C ratio is ≥ 1, choose the higher-cost alternative; otherwise, choose the lower-cost alternative.

B/C Ratio Criterion for Three or More Alternatives

Follow the logic for rate of return, except that the test is whether or not the incremental B/C ratio is ≥ 1.

EXAMPLE 23

Solve Example 22 using Benefit–Cost ratio analysis. Consider four mutually exclusive alternatives:

	Alternative			
	A	*B*	*C*	*D*
Initial Cost	$400.0	$100.0	$200.0	$500.0
Uniform Annual Benefit	100.9	27.7	46.2	125.2

Each alternative has a five-year useful life and no salvage value. Based on a 6% interest rate, which alternative should be selected?

Solution

1. B/C ratio computed for the alternatives:

Alt. *A* $B/C = \dfrac{\text{PW of benefits}}{\text{PW of cost}} = \dfrac{100.9(P/A,6\%,5)}{400} = 1.06$

B $B/C = \dfrac{27.7(P/A,6\%,5)}{100} = 1.17$

C $B/C = \dfrac{46.2(P/A,6\%,5)}{200} = 0.97$

D $B/C = \dfrac{125.2(P/A,6\%,5)}{500} = 1.05$

Alternative *C* with a B/C ratio less than 1 is eliminated.

2. Rank the remaining alternatives in order of increasing intial cost and examine the increment of investment between the two lowest cost alternatives.

Alternative	*B*	*A*	*D*
Initial Cost	$100.0	$400.0	$500.0
Uniform Annual Benefit	27.7	100.9	125.2

	A − B
Initial Cost	$300.0
Uniform Annual Benefit	73.2

Incremental B/C ratio $= \dfrac{73.2(P/A,6\%,5)}{300} = 1.03$

The incremental B/C ratio exceeds 1.0 hence the increment is desirable. Alternative *A* is preferred over *B*.

3. Do the next two-alternative comparison.

	Alternative		*Increment*
	A	*D*	*D − A*
Initial Cost	$400.0	$500.0	$100.0
Uniform Annual Benefit	100.9	125.2	24.3

Incremental B/C ratio $= \dfrac{24.3(P/A,6\%,5)}{100} = 1.02$

The incremental B/C ratio exceeds 1.0, hence Alternative D is preferred.

Conclusion: Select Alternative D.

BREAKEVEN ANALYSIS

In business, "breakeven" is defined as the point where income just covers the associated costs. In engineering economics, the breakeven point is more precisely defined as the point where two alternatives are equivalent.

EXAMPLE 24

A city is considering a new $50,000 snowplow. The new machine will operate at a savings of $600 per day, compared to the equipment presently being used. Assume the minimum attractive rate of return (interest rate) is 12% and the machine's life is 10 years with zero resale value at that time. How many days per year must the machine be used to make the investment economical?

Solution

This breakeven problem may be readily solved by annual cost computations. We will set the equivalent uniform annual cost of the snowplow equal to its annual benefit, and solve for the required annual utilization.

Let X = breakeven point = days of operation per year.

$$\text{EUAC} = \text{EUAB}$$
$$50,000(A/P,12\%,10) = 600X$$

$$X = \frac{50,000(0.1770)}{600} = 14.7 \text{ days/year}$$

DEPRECIATION

Depreciation of capital equipment is an important component of many after-tax economic analyses. For this reason, one must understand the fundamentals of depreciation accounting.

Depreciation is defined, in its accounting sense, as the systematic allocation of the cost of a capital asset over its useful life. *Book value* is defined as the original cost of an asset, minus the accumulated depreciation of the asset.

In computing a schedule of depreciation charges, three items are considered:

1. Cost of the property, P;

2. Depreciable life in years, n;

3. Salvage value of the property at the end of its depreciable life, S.

Straight Line Depreciation

Depreciation charge in any year $= \dfrac{P - S}{n}$

Sum-Of-Years-Digits Depreciation

$$\text{Depreciation charge} \atop \text{in any year} = \frac{\text{Remaining Depreciable Life at Beginning of Year}}{\text{Sum of Years Digits for Total Useful Life}}(P - S)$$

where Sum Of Years Digits $= 1 + 2 + 3 + \cdots + n = \frac{n}{2}(n + 1)$

Double Declining Balance Depreciation

Depreciation charge in any year $= \frac{2}{n}(P - \text{Depreciation charges to date})$

Accumulated Cost Recovery System (ACRS) Depreciation

ACRS depreciation is based on a property class life which is generally less than the actual useful life of the property and on zero salvage value. The varying depreciation percentage to use must be read from a table (based on declining balance with conversion to straight line). Unless one knows the proper ACRS property class, and has an ACRS depreciation table, the depreciation charge in any year cannot be computed.

EXAMPLE 25
A piece of machinery costs $5000 and has an anticipated $1000 salvage value at the end of its five-year depreciable life. Compute the depreciation schedule for the machinery by:

(a) Straight line depreciation;

(b) Sum-of-years-digits depreciation;

(c) Double declining balance depreciation.

Solution

Straight line depreciation $= \dfrac{P - S}{n} = \dfrac{5000 - 1000}{5} = \$800$

Sum-of-years-digits depreciation:

Sum-of-years-digits $= \frac{n}{2}(n + 1) = \frac{5}{2}(6) = 15$

1st year depreciation $= \frac{5}{15}(5000 - 1000) = \quad \$1333$

2nd year depreciation $= \frac{4}{15}(5000 - 1000) =$ 1067

3rd year depreciation $= \frac{3}{15}(5000 - 1000) =$ 800

4th year depreciation $= \frac{2}{15}(5000 - 1000) =$ 533

5th year depreciation $= \frac{1}{15}(5000 - 1000) =$ 267

                                                 $\overline{\$4000}$

Double declining balance depreciation:

1st year depreciation $= \frac{2}{5}(5000 - 0)$ $=$ $\$2000$

2nd year depreciation $= \frac{2}{5}(5000 - 2000)$ $=$ 1200

3rd year depreciation $= \frac{2}{5}(5000 - 3200)$ $=$ 720

4th year depreciation $= \frac{2}{5}(5000 - 3920)$ $=$ 432

5th year depreciation $= \frac{2}{5}(5000 - 4352)$ $=$ 259

                                                 $\overline{\$4611}$

Since the problem specifies a $1000 salvage value, the total depreciation may not exceed $4000. The double declining balance depreciation must be stopped in the 4th year when it totals $4000.

The depreciation schedules computed by the three methods are as follows:

Year	Straight Line	Sum-Of-Years-Digits	Double Declining Balance
1	$800	$1333	$2000
2	800	1067	1200
3	800	800	720
4	800	533	80
5	800	267	0
	$\overline{\$4000}$	$\overline{\$4000}$	$\overline{\$4000}$

INCOME TAXES

Income taxes represent another of the various kinds of disbursements encountered in an economic analysis. The starting point in an after-tax computation is the before-tax cash flow. Generally, the before-tax cash flow contains three types of entries:

1. Disbursements of money to purchase capital assets. These expenditures create no direct tax consequence for they are the exchange of one asset (cash) for another (capital equipment).

2. Periodic receipts and/or disbursements representing operating income and/or expenses. These increase or decrease the year-by-year tax liability of the firm.

3. Receipts of money from the sale of capital assets, usually in the form of a salvage value when the equipment is removed. The tax consequence depends on the relationship between the book value (cost − depreciation taken) of the asset and its salvage value.

Situation	Tax Consequence
Salvage value > Book value	Capital gain on difference
Salvage value = Book value	No tax consequence
Salvage value < Book value	Capital loss on difference

After the before-tax cash flow, the next step is to compute the depreciation schedule for any capital assets. Next, taxable income is the taxable component of the before-tax cash flow minus the depreciation. Then, the income tax is the taxable income times the appropriate tax rate. Finally, the after-tax cash flow is the before-tax cash flow adjusted for income taxes.

To organize these data, it is customary to arrange them in the form of a cash flow table, as follows:

Year	Before-tax cash flow	Depreciation	Taxable income	Income taxes	After-tax cash flow
0	.		.	.	.
1	.	.	.	.	.
.	.	.	.	.	.

EXAMPLE 26

A corporation expects to receive $32,000 each year for 15 years from the sale of a product. There will be an initial investment of $150,000. Manufacturing and sales expenses will be $8067 per year. Assume straight line depreciation, a 15-year useful life and no salvage value. Use a 46% income tax rate.

Determine the projected after-tax rate of return.

Solution

$$\text{Straight line depreciation} = \frac{P - S}{n} = \frac{150,000 - 0}{15}$$
$$= \$10,000 \text{ per year}$$

Year	Before-tax cash flow	Depreciation	Taxable income	Income taxes	After-tax cash flow
0	−150,000				−150,000
1	+23,933	10,000	13,933	−6,409	+17,524
2	+23,933	10,000	13,933	−6,409	+17,524
.	.	.	.	.	.
.	.	.	.	.	.
.	.	.	.	.	.
15	+23,933	10,000	13,933	−6,409	+17,524

Take the after-tax cash flow and compute the rate of return at which PW of cost equals PW of benefits.

$$150,000 = 17,524(P/A,i,15)$$

$$(P/A,i,15) = \frac{150,000}{17,524} = 8.559$$

From Compound Interest Tables, $i = 8\%$.

A CONCLUDING COMMENT

As you have seen, engineering economics is not a complex subject. There are, however, a group of fundamental concepts and techniques that must be mastered. This brief discussion has sought to provide a sound overview to these fundamentals. Unfortunately, there are likely to be other difficulties that you will encounter. In those situations, you will need to refer to one of the standard textbooks on the subject. I have listed my favorite reference (which should be no surprise) below.

Reference

Newnan, D. G. *Engineering Economic Analysis*, 3rd ed., 1988; Engineering Press, Inc., P.O. Box 1, San Jose, CA 95103-0001.

ENGR ECON 1

About how long will it take for $10,000 invested at 5% per year, compounded annually, to double in value?

(a)   5 yrs
(b) 10 yrs
(c) 15 yrs
(d) 20 yrs
(e) 25 yrs

P = $10,000   F = $20,000   i = 0.05   n = unknown

Using the single payment compound amount factor

$F = P(1+i)^n$ $1.05^n = \frac{20,000}{10,000} = 2$ n = 14.2 yrs ●

Alternate solution using compound interest tables

$F = P(F/P,5\%,n)$ $(F/P,5\%,n) = 20,000/10,000 = 2$
From 5% interest tables: $(F/P,5\%,14) = 1.98$ $(F/P,5\%,15) = 2.08$

n = 14.2 ●

Answer is (c)

ENGR ECON 2

If $200 is deposited in a savings account at the beginning of each of 15 years and the account draws interest at 7% per year, compounded annually, the value of the account at the end of 15 years will be most nearly

(a) $5000
(b) 5400
(c) 6000
(d) 6900
(e) 7200

$200

F' F

$$F' = A(F/A,i\%,n) = \$200(F/A,7\%,15) = 200(25.129) = \$5025.80$$
$$F = F'(F/P,i\%,n) = 5025.80(F/P,7\%,1) = 5025.80(1.07) = \$5377.61 \, \bullet$$

Answer is (b)

ENGR ECON 3

How many months at an interest rate of 1 percent per month does money have to be invested before it will double in value?

(a) 59 months
(b) 62
(c) 70
(d) 76
(e) 83

Let $P = \$1$ $\quad F = \$2$ $\quad i = 0.01$ per month $\quad n =$ number of months

$$F = P(1 + i)^n \qquad \$2 = \$1(1.01)^n \qquad 1.01^n = 2$$

The solution may be obtained from a 1% compound interest table or by hand calculator. $n = 70$ months. $\bullet$

Answer is (c)

ENGR ECON 4

A department store charges one and one-half percent interest per month on credit purchases. This is equivalent to a nominal annual interest rate of

(a) 1.5 percent
(b) 15.0
(c) 18.0
(d) 19.6
(e) 21.0

The nominal interest rate is the annual interest rate ignoring the effect of any compounding.

Nominal interest rate = $1\frac{1}{2}$ percent/month x 12 months = 18% ●

Answer is (c)

ENGR ECON 5

A bank charges $1\frac{1}{2}$% per month on the unpaid balance for purchases made with its credit card. This is equivalent to what effective annual interest rate?

(a) 1.5%
(b) 12%
(c) 18%
(d) 19.5%
(e) 39%

Effective interest rate is the annual interest rate, taking into account the effect of any compounding during the year.

Effective interest rate = $(1 + i)^m - 1$ where i = interest rate/interest period.
                                         m = number of compoundings per year.

$$= (1 + 0.015)^{12} - 1$$
$$= 0.1956 = 19.56\% ●$$

Answer is (d)

ENGR ECON 6

A bank pays one percent interest on savings accounts four times a year. The effective annual interest rate is

(a) 1.00%
(b) 1.04%
(c) 3.96%
(d) 4.00%
(e) 4.06%

Effective interest rate = $(1 + 0.01)^4 - 1 = 0.0406 = 4.06\%$ ●

Answer is (e)

ENGR ECON 7

What interest rate, compounded quarterly, is equivalent to a 9.31% effective interest rate?

(a) 2.25%
(b) 2.33%
(c) 4.66%
(d) 9.00%
(e) 9.31%

$0.0931 = (1 + i)^4 - 1$ $1.0931^{0.25} = 1 + i$
                                      $1.0225 = 1 + i$ i = 2.25% per quarter
                                                        i = 9% annual interest ●

Answer is (d)

ENGR ECON 8

A principal sum P is invested at a nominal interest rate r, compounded m times a year, for n years. The accumulated amount at the end of this period will be

(a) $P(1 + r/m)^r$

(b) $P(1 + r/m)^m$

(c) $P(1 + r/m)^{n/m}$

(d) $P(1 + r/m)^{m/r}$

(e)   None of these

The relationship between a future sum F and a present sum P is

$F = P(1 + i)^n$ where i = interest rate per interest period
                          n = number of interest periods

The nominal annual interest rate r is the sum of the individual interest payments in a one year period. Therefore, the interest rate per interest period i = r/m. The number of interest periods in a particular situation is the number of interest periods per year m times the number of years n. Note that the problem specifies n years, while conventional notation defines n as the number of interest periods. Thus the problem makes the number of interest periods equal to the number of compoundings per year m times the number of years.

$i = r/m$ and no. of interest periods = mn

Substituting these into

$F = P(1 + i)^n$ gives $F = P(1 + r/m)^{mn}$ ●

Answer is (e)

ENGR ECON 9

In the formula $P = F(1 + i)^{-n}$ the factor $(1 + i)^{-n}$ is called the

(a)   sinking fund factor
(b)   single payment present worth factor
(c)   single payment compound amount factor
(d)   capital recovery factor
(e)   uniform series present worth factor

Answer is (b) ●

ENGR ECON 10

A fund established to produce a desired amount at the end of a given period by means of a series of payments throughout the period is called a sinking fund, and is represented by the formula:

(a) $A = F \left[\dfrac{i}{(1 + i)^n - 1} \right]$

(d) $A = P \left[\dfrac{(1 + i)^n - 1}{i(1 + i)^n} \right]$

(b) $A = F \left[\dfrac{(1 + i)^n}{i} \right]$

(e) $A = F \left[\dfrac{i}{(1 + i)^n} \right]$

(c) $A = P(1 + i)^{-n}$

$$A = F \left[\frac{i}{(1 + i)^n - 1} \right] \bullet$$

Answer is (a)

ENGR ECON 11

Which of the following relationships between compound interest factors is NOT correct?

(a) Single payment compound amount factor and single payment present worth factor are reciprocals.

(b) Sinking fund factor and uniform series compound amount factor are reciprocals.

(c) Capital recovery factor and uniform series present worth factor are reciprocals.

(d) Capital recovery factor equals sinking fund factor plus the interest rate.

(e) Capital recovery factor and sinking fund factor are reciprocals.

Since answer (c) says Capital recovery = 1/(Series present worth)

and (e) says Capital recovery = 1/(Sinking fund)

it would necessarily follow that Series present worth = Sinking fund, and that obviously is not true. Thus the incorrect statement is either (c) or (e).

(a) $(F/P,i\%,n) = \dfrac{1}{(P/F,i\%,n)}$

(b) $(A/F,i\%,n) = \dfrac{1}{(F/A,i\%,n)}$

(c) $(A/P,i\%,n) = \dfrac{1}{(P/A,i\%,n)}$

(d) $(A/P,i\%,n) = (A/F,i\%,n) + i$

(e) $(A/P,i\%,n) \neq \dfrac{1}{(A/F,i\%,n)}$ $\bullet$

Answer is (e)

ENGR ECON 12

For some interest rate i and some number of interest periods n, the uniform series capital recovery factor is 0.0854 and the sinking fund factor is 0.0404. The interest rate i must be

(a) $3\frac{1}{2}\%$

(b) $4\frac{1}{2}\%$

(c) 6%

(d) 8%

(e) 9%

In Part (d) of ENGR ECON 11 we saw that

$$(A/P,i\%,n) = (A/F,i\%,n) + i$$

If we substitute the values given here into the equation, we have

$$0.0854 = 0.0404 + i$$
$$i = 0.0854 - 0.0404 = 0.045 = 4\tfrac{1}{2}\% \quad \bullet$$

Answer is (b)

ENGR ECON 13

An "annuity" is defined as

- (a) Earned interest due at the end of each interest period.
- (b) Cost of producing a product or rendering a service.
- (c) Total annual overhead assigned to a unit of production.
- (d) Amount of interest earned by a unit of principal in a unit of time.
- (e) A series of equal payments occurring at equal periods of time.

Answer is (e) ●

ENGR ECON 14

An individual wishes to deposit a certain quantity of money now so that at the end of five years he will have $500. With interest at 4% per year, compounded semiannually, how much must he deposit now?

- (a) $340.30
- (b) 400.00
- (c) 410.15
- (d) 416.95
- (e) 608.35

$P = F(1 + i)^{-n}$ where F = $500.

i = 0.02 per interest period (i is *not* 0.04)

n = 10 interest periods

Calculator solution:

$$P = \$500(1 + 0.02)^{-10} = \$410.17 \quad \bullet$$

Compound interest table solution:

$$P = \$500(P/F,2\%,10) = \$500(0.8203) = \$410.15 \quad \bullet$$

Answer is (c)

ENGR ECON 15

·The present worth of an obligation of $10,000 due in 10 years if money is worth 9% is nearest to

- (a) $10,000
- (b) 9,000
- (c) 7,500
- (d) 6,000
- (e) 4,500

$$P = F(P/F,9\%,10) = \$10,000(0.4224) = \$4224 \,\bullet$$

Answer is (e)

ENGR ECON 16

$1000 is borrowed for one year at an interest rate of 1% per month. If this same sum of money is borrowed for the same period at an interest rate of 12% per annum, the saving in interest charges would be:

(a) $ 0
(b)   3
(c)   5
(d)   7
(e)   14

Calculator solution:

At i = 1% per month $F = \$1000(1 + 0.01)^{12} = \$1126.83$
At i = 12% per year $F = 1000(1 + 0.12)^{1} = \underline{1120.00}$
Saving in interest = $6.83 •

Interest table solution:

At i = 1% per month $F = \$1000(F/P,1\%,12) = 1000(1.127) = \$1127.00$
At i = 12% per year $F = 1000(F/P,12\%,1) = 1000(1.120) = \underline{1120.00}$
Saving in interest = $7.00 •

Answer is (d)

ENGR ECON 17

A = $10,000

n = 15
i = 5%

F

F is closest to

(a) $105,000
(b)   150,000
(c)   157,000
(d)   215,000
(e)   262,000

$$F = \$10,000(F/A,5\%,15) = 10,000(21.579) = \$215,790 \,\bullet$$

Answer is (d)

ENGR ECON 18

A A A A A A A A A A A A

n = 12
i = 6%   F = $10,000

A is closest to

       (a) $ 593
       (b)   783
       (c)   833
       (d)   883
       (e)   1193

$$A = 10,000(A/F,6\%,12) = 10,000(0.0593) = \$593 \bullet$$

Answer is (a)

ENGR ECON 19

A company deposits $1000 every year for ten years in a bank. The company makes no deposits during the subsequent five years. If the bank pays 8% interest, the amount in the account at the end of 15 years is nearest to

       (a) $10,800
       (b)   15,000
       (c)   16,200
       (d)   21,200
       (e)   25,200

$$F = 1000(F/A,8\%,10)(F/P,8\%,5) = 1000(14.487)(1.469) = \$21,281.40 \bullet$$

Answer is (d)

ENGR ECON 20

$25,000 is deposited in a savings account that pays 5% interest, compounded semiannually. Equal annual withdrawals are to be made from the account, beginning one year from now and continuing forever. The maximum amount of the equal annual withdrawal is closest to:

       (a) $ 625
       (b)   1000
       (c)   1250
       (d)   1265
       (e)   1365

Effective interest rate $= (1 + 0.025)^2 - 1 = 0.050625 = 5.0625\%$

Annual withdrawal $A = Pi = 25,000(0.050625) = \$1265.63 \bullet$

Answer is (d)

ENGR ECON 21

What present sum would need to be put in a savings account now to provide a $1000 annual withdrawal for the next 50 years, if interest is 6%? The present sum is closest to:

       (a) $ 1,000
       (b)   10,000
       (c)   25,000
       (d)   37,500
       (e)   50,000

$$P = \$1000(P/A,6\%,50) = 1000(15.762) = \$15,762 \bullet$$

Answer is (b)

ENGR ECON 22

What present sum is equivalent to a series of $1000 annual end-of-year
payments, if a total of 10 payments are made and interest is 6%?
The present sum is closest to

(a) $ 6,250
(b)   7,350
(c)   9,400
(d)   10,000
(e)   10,600

P = 1000(P/A,6%,10) = 1000(7.360) = $7360 ●

Answer is (b)

ENGR ECON 23

A woman made ten annual end-of-year purchases of $1000 of common stock.
At the end of the tenth year she sold all the stock for $12,000.
What interest rate did she obtain on her investment?

(a)   2%
(b)   4%
(c)   8%
(d)   14%
(e)   20%

F = A(F/A,i%,n)   12,000 = 1000(F/A,i%,10)

$$(F/A,i\%,10) = \frac{12,000}{1,000} = 12$$

In the 4% interest table
(F/A,4%,10) = 12.006   i is very close to 4% ●

Answer is (b)

ENGR ECON 24

What present sum would be needed to provide for annual end-of-year payments
of $15 each, forever? Assume interest is 8%

(a) $120.00
(b)   137.50
(c)   150.00
(d)   187.50
(e)   375.00

In the special situation where n equals infinity, the value of the capital
recovery factor is i.

A = P(A/P,i%,∞) = Pi

$$P = \frac{A}{i} = \frac{\$15}{0.08} = \$187.50 ●$$

Answer is (d)

ENGR ECON 25

What amount of money deposited 50 years ago at 8% interest would now provide a
perpetual payment of $10,000 per year? The amount is nearest to:

(a) $ 3,000
(b)   8,000
(c)   50,000
(d)   82,000
(e)   125,000

The amount of money needed now to begin the perpetual payments:

$$P' = \frac{A}{i} = \frac{10,000}{0.08} = \$125,000$$

The amount of money that would need to have been deposited 50 years ago at 8% interest is:

$$P = \$125,000(P/F,8\%,50) = 125,000(0.0213) = \$2662.50 \bullet$$

Answer is (a)

ENGR ECON 26

A dam was constructed for $200,000. The annual maintenance cost is $5000. If interest is 5%, the capitalized cost of the dam, including maintenance, is:

    (a) $100,000
    (b)   200,000
    (c)   215,000
    (d)   250,000
    (e)   300,000

Capitalized cost is defined as the present worth of perpetual service, and is frequently used in connection with public works projects.

$$\text{Capitalized cost} = \$200,000 + \frac{A}{i} = \$200,000 + \frac{\$5000}{0.05} = \$300,000 \bullet$$

Answer is (e)

ENGR ECON 27

Given a sum of money Q that will be received six years from now. At 5% annual interest the present worth now of Q is $60. At this same interest rate, what would be the value of Q ten years from now?

    (a) $ 60.00
    (b)   76.58
    (c)   90.00
    (d)   97.74
    (e)   120.00

This problem illustrates the concept of equivalence. The present sum P = $60 is equivalent to Q six years hence at 5% annual interest.
The future sum F may be calculated by either of two methods:

$$F = Q(F/P,5\%,4) \qquad \text{Eqn (1)}$$
$$\text{or} \quad F = P(F/P,5\%,10) \qquad \text{Eqn (2)}$$

Since P is known, Eqn 2 may be solved directly.

$$F = P(F/P,5\%,10) = \$60(1.629) = \$97.74 \bullet$$

Answer is (d)

ENGR ECON 28

A piece of property is purchased for $10,000 and yields a $1000 yearly profit. If the property is sold after five years, what is the minimum price to break-even, with interest at 6%?

    (a) $ 5,000
    (b)   6,554
    (c)   7,743
    (d)   8,314
    (e)   10,000

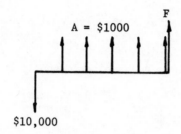

$$F = \$10,000(F/P,6\%,5) - \$1000(F/A,6\%,5)$$
$$= \$10,000(1.338) - \$1000(5.637) = 13,380 - 5637 = \$7743 \; \bullet$$

Answer is (c)

ENGR ECON 29

A steam boiler is purchased on the basis of guaranteed performance. A test indicates that the operating cost will be $300 more per year than the manufacturer guaranteed. If the expected life of the boiler is 20 years and money is worth 8%, how much should the purchaser deduct from the purchase price to compensate for the extra operating cost?

    (a) $2945
    (b)   3320
    (c)   4102
    (d)   5520
    (e)   6000

$$P = \$300(P/A,8\%,20) = 300(9.818) = \$2945.40 \; \bullet$$

Answer is (a)

ENGR ECON 30

Annual maintenance costs for a particular section of highway pavement are $2000. The placement of a new surface would reduce the annual maintenance cost to $500 per year for the first five years and to $1000 per year for the next five years. The annual maintenance after ten years would again be $2000. If maintenance costs are the only saving, what maximum investment can be justified for the new surface? Assume interest at 4%.

    (a) $ 5,500
    (b)   7,170
    (c)   10,000
    (d)   10,340
    (e)   12,500

Benefits = $1500 per year for the first five years and
            $1000 per year for the subsequent five years.
There are several ways of computing the present worth of benefits.
Two solutions will be presented.

Engineering Economics / 309

As the sketch indicates, the benefits may be considered as $1000 per year for ten years, plus an additional $500 benefit in each of the first five years.

Maximum investment = Present Worth of benefits
= $1000(P/A,4%,10) + $500(P/A,4%,5)
= $1000(8.111) + $500(4.452)
= $10,337 ●

Alternate Solution.

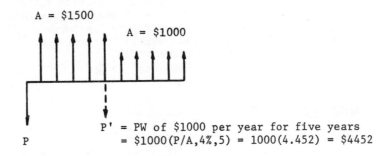

P' = PW of $1000 per year for five years
= $1000(P/A,4%,5) = 1000(4.452) = $4452

Maximum investment = Present Worth of Benefits
= $1500(P/A,4%,5) + P'(P/F,4%,5)
= $1500(4.452) + $4452(0.8219)
= $10,337 ●

Answer is (d)

ENGR ECON 31

A man buys a small garden tractor. There will be no maintenance cost the first year as the tractor is sold with one year's free maintenance. The second year the maintenance is estimated at $20. In subsequent years the maintenance cost will increase $20 per year (that is, 3rd year maintenance will be $40; 4th year maintenance will be $60, and so forth). How much would need to be set aside now at 5% interest to pay the maintenance costs on the tractor for the first six years of ownership?

(a) $101.52
(b) 164.74
(c) 239.36
(d) 284.13
(e) 300.00

Solution using single payment present worth factors

$$P = 20(P/F,5\%,2) + 40(P/F,5\%,3) + 60(P/F,5\%,4) + 80(P/F,5\%,5) + 100(P/F,5\%,6)$$
$$= 20(0.9070) + 40(0.8638) + 60(0.8227) + 80(0.7835) + 100(0.7462)$$
$$= \$239.35 \; \bullet$$

Alternate solution using the gradient present worth factor

$$P = 20(P/G,5\%,6) = 20(11.968) = \$239.36 \; \bullet$$

Answer is (c)

ENGR ECON 32

Which one of the following is *NOT* a method of depreciating plant equipment for accounting and engineering economic analysis purposes?

 (a) Double entry method
 (b) Fixed percentage method
 (c) Sum-of-years digits method
 (d) Straight line method
 (e) Sinking fund method

Answer is (a) $\bullet$

ENGR ECON 33

A manufacturing company buys a machine for $50,000. It estimates the machine's useful life is 20 years and that it can then be sold for $5000. Using straight line depreciation, what is the annual depreciation charge?

 (a) $2000
 (b) 2250
 (c) 2500
 (d) 2750
 (e) 3000

Straight line depreciation

$$\text{Annual depreciation charge} = \frac{P - S}{n} = \frac{50,000 - 5,000}{20} = \$2250 \; \bullet$$

Answer is (b)

ENGR ECON 34

In determining the average annual cost of a proposed project, the formula

$$\frac{P - S}{n} + (P - S)\left(\frac{i}{2}\right)\left(\frac{n + 1}{n}\right) + Si \quad \text{represents the economic method of}$$

 (a) Sinking fund depreciation plus interest on first cost.

 (b) Straight line depreciation plus interest on first cost.

 (c) Straight line depreciation plus average interest.

 (d) Capital recovery with a return.

 (e) Amortization plus interest on first cost.

Straight line depreciation $= \dfrac{P - S}{n}$

Average interest $= \dfrac{\text{First year's interest} + \text{Last year's interest}}{2}$

First year's interest $= (P - S)i + Si$

Last year's interest $= \dfrac{(P - S)}{n}(i) + Si$

Average interest $= \dfrac{(P - S)i + Si + \dfrac{(P - S)}{n}(i) + Si}{2}$

$= (P - S)\left(\dfrac{i}{2}\right)\left(\dfrac{n + 1}{n}\right) + Si$

Therefore, straight line depreciation plus average interest equals

$$\dfrac{P - S}{n} + (P - S)\left(\dfrac{i}{2}\right)\left(\dfrac{n + 1}{n}\right) + Si \quad \bullet$$

Answer is (c)

ENGR ECON 35

Company A has fixed expenses of $15,000 per year and each unit of product has a $0.002 variable cost. Company B has fixed expenses of $5000 per year and can produce the same unit of product at a $0.05 variable cost. At what number of units of annual production will Company A have the same overall cost as Company B? Quantity is nearest to

  (a)   100,000 units
  (b)   200,000
  (c)   300,000
  (d)   400,000
  (e) 2,000,000

Let x = annual production (units)

Total cost to Company A = Total cost to Company B

$$15,000 + 0.002x = 5000 + 0.05x$$

$$x = \dfrac{10,000}{0.048} = 208,333 \text{ units} \quad \bullet$$

Answer is (b)

ENGR ECON 36

Plan A requires a $100,000 investment now.
Plan B requires an $80,000 investment now and an additional $40,000 investment later. At 8% interest, what is the breakeven point on the timing of the additional $40,000 later?

  (a)   3 years
  (b)   5
  (c)   7
  (d)   9
  (e) 11

The difference between the alternatives is that Plan A requires $20,000 extra now and Plan B requires $40,000 extra n years hence.

At breakeven

$$\$20,000 = \$40,000(P/F,8\%,n)$$

$$(P/F,8\%,n) = 0.5$$

From the 8% compound interest table $(P/F,8\%,9) = 0.5002$
Therefore, n = 9 years. ●

Answer is (d)

ENGR ECON 37   *(This is a 10-question problem)*

A company is considering buying a new piece of machinery. Two models are available.

	Machine I	Machine II
Initial cost	$80,000	$100,000
End of useful life salvage value, S	20,000	25,000
Annual operating cost	18,000	15,000 first 10 years; 20,000 thereafter
Useful life	20 yrs	25 yrs

Answer the following ten questions, based on a 10% interest rate.

QUESTION 1 What is the equivalent uniform annual cost for Machine I?
The annual cost is closest to
        (a) $21,000
        (b)   23,000
        (c)   25,000
        (d)   27,000
        (e)   29,000

EUAC = (P - S)(A/P,i%,n) + Si + Annual operating cost

  = (80,000 - 20,000)(A/P,10%,20) + 20,000(0.10) + 18,000

  = 60,000(0.1175) + 2000 + 18,000

  = 27,050 ●

Answer is (d)

QUESTION 2 What is the equivalent uniform annual cost for Machine II?
The annual cost is closest to
        (a) $21,000
        (b)   23,000
        (c)   25,000
        (d)   27,000
        (e)   29,000

EUAC = (100,000 - 25,000)(A/P,10%,25) + 25,000(0.10)

   + 20,000 - 5000(P/A,10%,10)(A/P,10%,25)

  = 75,000(0.1102) + 2500 + 20,000 - 5000(6.145)(0.1102)

  = $27,379 ●

Answer is (d)

QUESTION 3 The capitalized cost based on Machine I is closest to
  (a) $ 60,000
  (b)   80,000
  (c)   230,000
  (d)   270,000
  (e)   420,000

Capitalized cost = Present worth of an infinite life.
In Question 1 we computed the Equivalent Uniform Annual Cost (EUAC).

$$P = \frac{A}{i}$$

Capitalized Cost $= \frac{A}{i} = \frac{EUAC}{i} = \frac{\$27,050}{0.10} = \$270,500$ ●

Answer is (d)

QUESTION 4 If Machine I is purchased and a fund is set up to replace Machine I
at the end of 20 years, the uniform annual deposit that should be
made to the fund is nearest to:
  (a) $1000
  (b)   2000
  (c)   3000
  (d)   4000
  (e)   5000

Required future sum F = $80,000 - 20,000 = $60,000

Annual deposit A = $60,000(A/F,10%,20) = $60,000(0.0175) = $1050 ●

Answer is (a)

QUESTION 5 Machine I will produce an annual savings in material of $25,700 a
year. What is the before-tax rate of return if Machine I is
installed? The rate of return is closest to:
  (a)   6%
  (b)   8%
  (c) 10%
  (d) 20%
  (e) 35%

The cash flow for this situation is:

Year	Cash Flow
0	-$80,000
1-20	$\begin{cases} +25,700 \\ -18,000 \end{cases}$
20	+20,000

Write one equation with i as the only unknown.

80,000 = (25,700 - 18,000)(P/A,i%,20) + 20,000(P/F,i%,20)

Try i = 8%

80,000 = 7700(9.818) + 20,000(0.2145) = 79,888

Therefore, the rate of return is very close to 8% ●

Answer is (b)

QUESTION 6 Assuming Sum-Of-Years Digits depreciation, what would be the book value of Machine I after two years? Book value is closest to:

    (a) $21,000
    (b)   42,000
    (c)   59,000
    (d)   69,000
    (e)   79,000

Sum-Of-Years Digits depreciation:

$$\text{Depreciation charge in any year} = \frac{\text{Remaining useful life at beginning of year}}{\text{Sum-Of-Years Digits for total useful life}}(P - S)$$

$$\text{Sum-Of-Years Digits} = \frac{n}{2}(n + 1) = \frac{20}{2}(21) = 210$$

$$\text{1st year depreciation} = \frac{20}{210}(80,000 - 20,000) = \$\ 5,714$$

$$\text{2nd year depreciation} = \frac{19}{210}(80,000 - 20,000) = \underline{\quad 5,429}$$
$$\text{TOTAL} = \$11,143$$

Book Value = Cost - Depreciation to date
        = $80,000 - 11,143
        = $68,857 ●

        Answer is (d)

QUESTION 7 Assuming Double Declining Balance depreciation, what would be the book value of Machine II after 3 years?

    (a) $16,000
    (b)   22,000
    (c)   58,000
    (d)   78,000
    (e)   83,000

Double Declining Balance depreciation:

$$\text{Depreciation charge in any year} = \frac{2}{n}(P - \text{Depreciation charges to date})$$

$$\text{1st year depreciation} = \frac{2}{25}(100,000 - 0) \qquad = \$8,000$$

$$\text{2nd year depreciation} = \frac{2}{25}(100,000 - 8,000) = 7,360$$

$$\text{3rd year depreciation} = \frac{2}{25}(100,000 - 15,360) = \underline{\quad 6,771}$$
$$\text{TOTAL: } \$22,131$$

Book Value = Cost - Depreciation to date
        = $10,000 - 22,131
        = $77,869 ●

        Answer is (d)

QUESTION 8 A new building is being considered to house some equipment. The new building will reduce maintenance costs by $6000 per year for the first ten years, and $3000 per year thereafter. Based on a 50-year analysis period, what building construction cost can be justified? The justified construction cost is closest to:

(a) $ 30,000
(b)   50,000
(c)   90,000
(d) 140,000
(e) 180,000

The justified construction cost = Present Worth of savings

P = $6000(P/A,10%,10) + $3000(P/A,10%,40)(P/F,10%,10)

  = $6000(6.145) + $3000(9.779)(0.3855)

  = $48,179 ●

Alternate solution:

P = $3000(P/A,10%,50) + $3000(P/A,10%,10)

  = $3000(9.915) + $3000(6.145)

  = $48,180 ●

       Answer is (b)

QUESTION 9 The manufacturers of Machine II have announced a price reduction for the machine. What is the breakeven initial price for Machine II, when compared to Machine I?

(a) $ 91,000
(b)   94,000
(c)   97,000
(d) 100,000
(e) 103,000

In *Question 1* the equivalent uniform annual cost for Machine I was computed to be $27,050.
For breakeven set the EUAC of Machine I equal to the EUAC of Machine II, and compute initial cost of Machine II at this point.

$EUAC_I = EUAC_{II}$

$27,050 = (P - 25,000)(A/P,10%,25) + 25,000(0.10) + 20,000$
       $- 5000(P/A,10%,10)(A/P,10%,25)$

    $= (P - 25,000)(0.1102) + 22,500 - 5000(6.145)(0.1102)$

    $= 0.1102P - 2755 + 22,500 - 3386$

$$\$27,050 = 0.1102P + 16,359$$

$$P = \frac{\$27,050 - 16,359}{0.1102} = \$97,015 \; \bullet$$

Answer is (c)

QUESTION 10 What must be the salvage value of Machine I at the end of 20 years for the machine to have an equivalent uniform annual cost of $26,500?

(a) $10,000
(b) 20,000
(c) 30,000
(d) 40,000
(e) 50,000

We can write one equation with S as the only unknown.

$$\$26,500 = \$80,000(A/P,10\%,20) + 18,000 - S(A/F,10\%,20)$$

$$= \$80,000(0.1175) + 18,000 - S(0.0175)$$

$$= \$9400 + 18,000 - 0.0175S$$

$$S = \frac{\$27,400 - \$26,500}{0.0175} = \$51,429 \; \bullet$$

Answer is (e)

ENGR ECON 38

A company uses 8000 wheels per year in its manufacture of golf carts. The wheels cost $15 each and are purchased from an outside supplier. The money invested in the inventory costs 10% per year, and the warehousing cost amounts to an additional 2% per year. It costs $150 to process each purchase order.

The quantity of wheels that should be ordered each time an order is placed is nearest to

(a) 1200
(b) 1600
(c) 2000
(d) 2500
(e) 3000

The simplest model for the Economic Order Quantity is

$$EOQ = \sqrt{\frac{2BD}{E}} \qquad \text{where} \qquad$$

B = Ordering cost, $/order
D = Demand per period, units
E = Inventory holding cost, $/unit/period
EOQ = Economic order quantity, units

$$EOQ = \sqrt{\frac{2 \times \$150 \times 8000}{(10\% + 2\%)(15.00)}} = 1155 \text{ wheels} \bullet$$

Answer is (a)

ENGR ECON 39

A bank pays 10% nominal annual interest on special three year certificates. Answer the following three questions.

QUESTION 1 If the interest is compounded every three months, the effective annual interest rate is nearest to

  (a) 10.000%
  (b) 10.375
  (c) 10.500
  (d) 10.750
  (e) 30.000

Effective $i = (1 + i)^m - 1 = (1 + 0.025)^4 - 1$

$$= 0.1038 = 10.38\% \bullet$$

Answer is (b)

QUESTION 2 If interest is compounded daily, the effective annual interest rate is nearest to

  (a) 10.000%
  (b) 10.375
  (c) 10.500
  (d) 10.750
  (e) 30.000

Effective $i = (1 + i)^m - 1 = \left[1 + \dfrac{0.10}{365}\right]^{365} - 1$

$$= 0.10516 = 10.516\% \bullet$$

Answer is (c)

QUESTION 3 If interest is compounded continuously, the effective annual interest rate is nearest to

  (a) 10.000%
  (b) 10.375
  (c) 10.500
  (d) 10.750
  (e) 30.000

For continuous compounding

Effective $i = e^r - 1$ where r = nominal interest rate

$= e^{0.10} - 1 = 0.10517 = 10.517\%$ ●

Answer is (c)

ENGR ECON 40

Calculate the rate of return for the following investment opportunity.

Invest $100 now.

Receive two payments of $109.46 - one at the end of year 3 and one at the end of year 6.

The rate of return is nearest to

(a) 10%
(b) 20
(c) 30
(d) 40
(e) 50

Set the Present Worth of cost = Present Worth of benefits

$$100 = 109.46(P/F,i\%,3) + 109.46(P/F,i\%,6)$$

Solve by trial and error:

Try i = 15%

$$100 \overset{?}{=} 109.46(0.6575) + 109.46(0.4323)$$

$$\overset{?}{=} 109.46(1.0898) = 119.29$$

The PW of benefits is greater than the PW of cost, indicating that the interest rate i is too low.

Try i = 20%

$$100 \overset{?}{=} 109.46(0.5787) + 109.46(0.3349)$$

$$\overset{?}{=} 109.46(0.9136) = 100.00$$

Here the PW of cost = PW of benefits when i = 20%.

Therefore, the rate of return = 20% ●

In most situations a trail i will not result in PW of Cost equal to the PW of benefits. A linear interpolation for i is usually required.

Answer is (b)

ENGR ECON 41

A machine part, operating in a corrosive atmosphere, is made of low-carbon steel, costs $350 installed and lasts six years. If the part is treated for corrosion resistance it will cost $700 installed.

How long must the treated part last to be as good an investment as the untreated part, if money is worth 7%?

(a) 8 years
(b) 11
(c) 14
(d) 17
(e) 20

The Equivalent Uniform Annual Cost ("annual cost") of the untreated part is:

$$\$350(A/P,7\%,6) = 350(0.2098) = \$73.43$$

For breakeven, the annual cost of the treated part must be this same value, or

$$\$73.43 = \$700(A/P,7\%,n)$$

$$(A/P,7\%,n) = \frac{73.43}{700} = 0.1049$$

From compound interest tables:

n	(A/P,7%,n)
16 yrs	0.1059
17 yrs	0.1024

By linear interpolation we get

$$n = 16 + (1) \frac{0.1059 - 0.1049}{0.1059 - 0.1024} = 16 + \frac{0.0010}{0.0035} = 16.3 \text{ years}$$

Thus the treated part must last 16.3 years to be as good an investment as the untreated part, that is, breakeven is 16.3 years. ●

Answer is (d)

ENGR ECON 42

An investor is considering buying a 20-year corporate bond. The bond has a face value of $1000 and pays 6% interest per year in two semiannual payments. Thus the purchaser of the bond would receive $30 every 6 months, and in addition he would receive $1000 at the end of 20 years, along with the last $30 interest payment.

If the investor though he should receive 8% annual interest, compounded semiannually, the amount he would be willing to pay for the bond is closest to

(a) $500
(b) $600
(c) $700
(d) $800
(e) $900

The investor would be willing to pay the present worth of the future benefits, computed at 8% compounded semiannually.

$1000

A = $30

P

For semiannual interest periods
$$i = 4\% \text{ per interest period}$$
$$n = 40 \text{ interest periods}$$

Present Worth $= 30(P/A,4\%,40) + 1000(P/F,4\%,40)$

$= 30(19.793) + 1000(0.2083)$

$= 593.79 + 208.30$

$= \underline{\$802.09}$ ●

Answer is (d)

ENGR ECON 43

An engineer is faced with the prospect of a fluctuating future budget for the maintenance of a particular machine. During each of the first five years, $1000 per year will be budgeted. During the second five years, the annual budget will be $1500 per year. In addition, $3500 will be budgeted for an overhaul of the machine at the end of the fourth year, and another $3500 for an overhaul at the end of the eighth year.

The engineer wonders what uniform annual expenditure would be equivalent to these fluctuating amounts, assuming compound interest at 6% per year.

The equivalent uniform annual expenditure for the 10-year period is closest to

    (a) $1050
    (b) $1250
    (c) $1450
    (d) $1650
    (e) $1850

A diagram of the projected budget amounts is as follows:

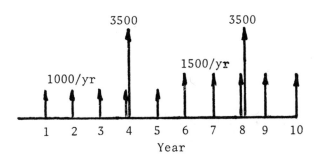

3500   3500

1500/yr

1000/yr

1 2 3 4 5 6 7 8 9 10
Year

Compute the present worth of the various future amounts.
Then compute the equivalent uniform annual amount.

PW = 1000(P/A,6%,5) + 3500(P/F,6%,4) + 1500(P/A,6%,5)(P/F,6%,5)
    + 3500(P/F,6%,8)

  = 1000(4.212) + 3500(0.7921) + 1500(4.212)(0.7473)
    + 3500(0.6274)

  = 4212 + 2772 + 4721 + 2196 = $13,901

Equivalent uniform annual amount = 13,901(A/P,6%,10)

          = 13,901(0.1359) = $1889 ●

Answer is (e)

ENGR ECON 44

A trust fund is to be established to
    (a) Provide $750,000 for the construction and $250,000
        for the initial equipment of an Electrical Engineering
        Laboratory.
    (b) Pay the laboratory annual operating costs of
        $150,000 per year.
    (c) Pay for $100,000 of replacement equipment every four
        years, beginning four years from now.
At 4% interest, how much money would be required in the trust fund
to provide for the laboratory and equipment and its perpetual
operation and equipment replacement?

  The amount is closest to
        (a) $1,000,000
        (b) $3,750,000
        (c) $4,750,000
        (d) $5,250,000
        (e) $7,500,000

Required money in trust fund = $1,000,000 initial construction and
equipment money
+ money for annual operation
+ money for periodic equipment
replacement.

For n = ∞

$$A = Pi \quad \text{or } P = \frac{A}{i}$$

For (a) P = $1,000,000

(b) $P = \frac{A}{i} = \frac{\$150,000}{0.04} = \$3,750,000$

(c) This situation is more complex. We do not know
what equivalent amount A will provide for the necessary
replacement equipment.

The perpetual series:

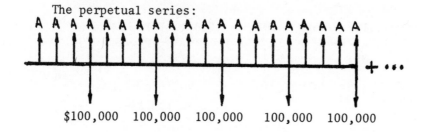

$100,000   100,000   100,000   100,000   100,000

We can solve one portion of the perpetual series for A:

$100,000

A = $100,000(A/F,4%,4) = $100,000(0.2355) = $23,550

This value of A computed for the four year period is the same
as the value of A for the perpetual series.

For the perpetual series:

$$P = \frac{A}{i} = \frac{\$23,550}{0.04} = \$589,000$$

Required money in trust fund = $1,000,000 + 3,750,000 + 589,000

= $5,339,000. ●

Answer is (d)

Appendix A

Examination Topics

The National Council of Engineering Examiners provides a detailed listing of the typical content of each topic.

MORNING SESSION
Mathematics
    Analytic Geometry
    Differential Calculus
    Integral Calculus
    Differential Equations
    Probability and Statistics
    Linear Algebra

Computer Programming
    Number Systems
    Algorithms
    Computer Organization
    Problem Solving with FORTRAN and BASIC

Statics
    Vector Algebra
    Two Dimensional Equilibrium
    Three Dimensional Equilibrium
    Concurrent Force Systems
    Centroid of Area
    Moment of Inertia
    Friction

Dynamics
    Kinematics
    Force, Mass, and Acceleration
    Impulse and Momentum
    Work and Energy

Mechanics of Materials
    Stress and Strain
    Tension
    Shear
    Beams
    Columns
    Torsion
    Bending

Fluid Mechanics
    Fluid Properties
    Fluid Statics
    Pipe Flow
    Flow Measurement
    Impulse and Momentum
    Similitude
    Dimensional Analysis

Thermodynamics
    Properties: Enthalpy, Entropy, Free Energy
    Thermodynamic Processes
    Energy, Heat, and Work
    Phase Changes
    First Law
    Second Law
    Availability-Reversibility
    Cycles
    Ideal Gases
    Mixture of Gases

Chemistry
    Nomenclature
    Equations
    Stoichiometry
    Periodicity
    States of Matter
    Metals and Nonmetals
    Oxidation and Reduction
    Acids and Bases
    Solutions
    Kinetics
    Equilibrium
    Electrochemistry
    Organic Chemistry

Materials Science
    Physical Properties
    Chemical Properties
    Electrical Properties
    Atomic Bonding
    Crystallography
    Phase Diagrams
    Processing Tests and Properties
    Diffusion
    Corrosion

Structure of Matter
    Nuclei
    Atoms
    Molecules

Electrical Circuits
 DC Circuits
 AC Circuits
 Three-phase Circuits
 Capacitance and Inductance
 Electric and Magnetic Fields

Engineering Economics
 Time Value of Money
 Annual Cost
 Present Worth
 Future Worth
 Capitalized Cost
 Breakeven Analysis
 Valuation and Depreciation

AFTERNOON SESSION
Required Topics:

Engineering Mechanics
 Statics
  Resultants of Force Systems
  Equilibria of Rigid Bodies
  Analysis of Internal Forces
  Couples and Moments
 Dynamics
  Kinematics
   Relative Motion
   Curvilinear Motion
  Inertia and Force
  Particles and Solid Bodies
  Impulse and Momentum
  Work and Energy
 Mechanics of Materials
  Stress and Strain
  Tension
  Shear
  Simple Beam
  Torsion
  Bending

Mathematics
 Same as morning session

Electrical Circuits
 DC Circuits
 AC Circuits
 Three-phase Circuits
 Capacitance and Inductance
 Electric and Magnetic Fields
 Transients

Engineering Economics
    Same as morning session

Optional Topics:

Computer Programming
    Same as morning session, but with emphasis on FORTRAN
    and BASIC

Electronics and Electrical Machinery
    Solid-state Devices and Circuits
    Electrical Machinery

Fluid Mechanics
    Fluid Statics
    Pipe Flow
    Compressible Flow
    Flow Measurement
    Impulse and Momentum
    Similitude
    Hydraulics

Mechanics of Materials
    Stress and Strain
    Tension
    Shear
    Combined Stress
    Beams
    Columns
    Composite Sections
    Uniaxial Loading
    Torsion
    Bending
    Combined Loading
    Shear and Moment Diagrams

Thermodynamics/Heat Transfer
    Chemical Reactions
    Cycles
        Thermal
        Gas Power
        Vapor Power
    Refrigeration
    Air Conditioning
    Combustion
    Gas Mixtures
    Flow through Nozzles, Turbines, and Compressors
    Availability and Reversibility
    Heat Transfer

Compound Interest Tables

Values of Interest Factors When *N* Equals Infinity

Single Payment:

$$(F/P,i,\infty) = \infty$$

$$(P/F,i,\infty) = 0$$

Arithmetic Gradient Series:

$$(A/G,i,\infty) = 1/i$$

$$(P/G,i,\infty) = 1/i^2$$

Uniform Payment Series:

$$(A/F,i,\infty) = 0$$

$$(A/P,i,\infty) = i$$

$$(F/A,i,\infty) = \infty$$

$$(P/A,i,\infty) = 1/i$$

½%

| | Single Payment | | Uniform Payment Series | | | | Arithmetic Gradient | | |
	Compound Amount Factor Find F Given P F/P	Present Worth Factor Find P Given F P/F	Sinking Fund Factor Find A Given F A/F	Capital Recovery Factor Find A Given P A/P	Compound Amount Factor Find F Given A F/A	Present Worth Factor Find P Given A P/A	Gradient Uniform Series Find A Given G A/G	Gradient Present Worth Find P Given G P/G	n
1	1.005	.9950	1.0000	1.0050	1.000	0.995	0	0	1
2	1.010	.9901	.4988	.5038	2.005	1.985	0.499	0.991	2
3	1.015	.9851	.3317	.3367	3.015	2.970	0.996	2.959	3
4	1.020	.9802	.2481	.2531	4.030	3.951	1.494	5.903	4
5	1.025	.9754	.1980	.2030	5.050	4.926	1.990	9.803	5
6	1.030	.9705	.1646	.1696	6.076	5.896	2.486	14.660	6
7	1.036	.9657	.1407	.1457	7.106	6.862	2.980	20.448	7
8	1.041	.9609	.1228	.1278	8.141	7.823	3.474	27.178	8
9	1.046	.9561	.1089	.1139	9.182	8.779	3.967	34.825	9
10	1.051	.9513	.0978	.1028	10.228	9.730	4.459	43.389	10
11	1.056	.9466	.0887	.0937	11.279	10.677	4.950	52.855	11
12	1.062	.9419	.0811	.0861	12.336	11.619	5.441	63.218	12
13	1.067	.9372	.0746	.0796	13.397	12.556	5.931	74.465	13
14	1.072	.9326	.0691	.0741	14.464	13.489	6.419	86.590	14
15	1.078	.9279	.0644	.0694	15.537	14.417	6.907	99.574	15
16	1.083	.9233	.0602	.0652	16.614	15.340	7.394	113.427	16
17	1.088	.9187	.0565	.0615	17.697	16.259	7.880	128.125	17
18	1.094	.9141	.0532	.0582	18.786	17.173	8.366	143.668	18
19	1.099	.9096	.0503	.0553	19.880	18.082	8.850	160.037	19
20	1.105	.9051	.0477	.0527	20.979	18.987	9.334	177.237	20
21	1.110	.9006	.0453	.0503	22.084	19.888	9.817	195.245	21
22	1.116	.8961	.0431	.0481	23.194	20.784	10.300	214.070	22
23	1.122	.8916	.0411	.0461	24.310	21.676	10.781	233.680	23
24	1.127	.8872	.0393	.0443	25.432	22.563	11.261	254.088	24
25	1.133	.8828	.0377	.0427	26.559	23.446	11.741	275.273	25
26	1.138	.8784	.0361	.0411	27.692	24.324	12.220	297.233	26
27	1.144	.8740	.0347	.0397	28.830	25.198	12.698	319.955	27
28	1.150	.8697	.0334	.0384	29.975	26.068	13.175	343.439	28
29	1.156	.8653	.0321	.0371	31.124	26.933	13.651	367.672	29
30	1.161	.8610	.0310	.0360	32.280	27.794	14.127	392.640	30
36	1.197	.8356	.0254	.0304	39.336	32.871	16.962	557.564	36
40	1.221	.8191	.0226	.0276	44.159	36.172	18.836	681.341	40
48	1.270	.7871	.0185	.0235	54.098	42.580	22.544	959.928	48
50	1.283	.7793	.0177	.0227	56.645	44.143	23.463	1 035.70	50
52	1.296	.7716	.0169	.0219	59.218	45.690	24.378	1 113.82	52
60	1.349	.7414	.0143	.0193	69.770	51.726	28.007	1 448.65	60
70	1.418	.7053	.0120	.0170	83.566	58.939	32.468	1 913.65	70
72	1.432	.6983	.0116	.0166	86.409	60.340	33.351	2 012.35	72
80	1.490	.6710	.0102	.0152	98.068	65.802	36.848	2 424.65	80
84	1.520	.6577	.00961	.0146	104.074	68.453	38.576	2 640.67	84
90	1.567	.6383	.00883	.0138	113.311	72.331	41.145	2 976.08	90
96	1.614	.6195	.00814	.0131	122.829	76.095	43.685	3 324.19	96
100	1.647	.6073	.00773	.0127	129.334	78.543	45.361	3 562.80	100
104	1.680	.5953	.00735	.0124	135.970	80.942	47.025	3 806.29	104
120	1.819	.5496	.00610	.0111	163.880	90.074	53.551	4 823.52	120
240	3.310	.3021	.00216	.00716	462.041	139.581	96.113	13 415.56	240
360	6.023	.1660	.00100	.00600	1 004.5	166.792	128.324	21 403.32	360
480	10.957	.0913	.00050	.00550	1 991.5	181.748	151.795	27 588.37	480

1%

| | Single Payment | | Uniform Payment Series | | | | Arithmetic Gradient | | |
n	Compound Amount Factor Find F Given P F/P	Present Worth Factor Find P Given F P/F	Sinking Fund Factor Find A Given F A/F	Capital Recovery Factor Find A Given P A/P	Compound Amount Factor Find F Given A F/A	Present Worth Factor Find P Given A P/A	Gradient Uniform Series Find A Given G A/G	Gradient Present Worth Find P Given G P/G	
1	1.010	.9901	1.0000	1.0100	1.000	0.990	0	0	1
2	1.020	.9803	.4975	.5075	2.010	1.970	0.498	0.980	2
3	1.030	.9706	.3300	.3400	3.030	2.941	0.993	2.921	3
4	1.041	.9610	.2463	.2563	4.060	3.902	1.488	5.804	4
5	1.051	.9515	.1960	.2060	5.101	4.853	1.980	9.610	5
6	1.062	.9420	.1625	.1725	6.152	5.795	2.471	14.320	6
7	1.072	.9327	.1386	.1486	7.214	6.728	2.960	19.917	7
8	1.083	.9235	.1207	.1307	8.286	7.652	3.448	26.381	8
9	1.094	.9143	.1067	.1167	9.369	8.566	3.934	33.695	9
10	1.105	.9053	.0956	.1056	10.462	9.471	4.418	41.843	10
11	1.116	.8963	.0865	.0965	11.567	10.368	4.900	50.806	11
12	1.127	.8874	.0788	.0888	12.682	11.255	5.381	60.568	12
13	1.138	.8787	.0724	.0824	13.809	12.134	5.861	71.112	13
14	1.149	.8700	.0669	.0769	14.947	13.004	6.338	82.422	14
15	1.161	.8613	.0621	.0721	16.097	13.865	6.814	94.481	15
16	1.173	.8528	.0579	.0679	17.258	14.718	7.289	107.273	16
17	1.184	.8444	.0543	.0643	18.430	15.562	7.761	120.783	17
18	1.196	.8360	.0510	.0610	19.615	16.398	8.232	134.995	18
19	1.208	.8277	.0481	.0581	20.811	17.226	8.702	149.895	19
20	1.220	.8195	.0454	.0554	22.019	18.046	9.169	165.465	20
21	1.232	.8114	.0430	.0530	23.239	18.857	9.635	181.694	21
22	1.245	.8034	.0409	.0509	24.472	19.660	10.100	198.565	22
23	1.257	.7954	.0389	.0489	25.716	20.456	10.563	216.065	23
24	1.270	.7876	.0371	.0471	26.973	21.243	11.024	234.179	24
25	1.282	.7798	.0354	.0454	28.243	22.023	11.483	252.892	25
26	1.295	.7720	.0339	.0439	29.526	22.795	11.941	272.195	26
27	1.308	.7644	.0324	.0424	30.821	23.560	12.397	292.069	27
28	1.321	.7568	.0311	.0411	32.129	24.316	12.852	312.504	28
29	1.335	.7493	.0299	.0399	33.450	25.066	13.304	333.486	29
30	1.348	.7419	.0287	.0387	34.785	25.808	13.756	355.001	30
36	1.431	.6989	.0232	.0332	43.077	30.107	16.428	494.620	36
40	1.489	.6717	.0205	.0305	48.886	32.835	18.178	596.854	40
48	1.612	.6203	.0163	.0263	61.223	37.974	21.598	820.144	48
50	1.645	.6080	.0155	.0255	64.463	39.196	22.436	879.417	50
52	1.678	.5961	.0148	.0248	67.769	40.394	23.269	939.916	52
60	1.817	.5504	.0122	.0222	81.670	44.955	26.533	1 192.80	60
70	2.007	.4983	.00993	.0199	100.676	50.168	30.470	1 528.64	70
72	2.047	.4885	.00955	.0196	104.710	51.150	31.239	1 597.86	72
80	2.217	.4511	.00822	.0182	121.671	54.888	34.249	1 879.87	80
84	2.307	.4335	.00765	.0177	130.672	56.648	35.717	2 023.31	84
90	2.449	.4084	.00690	.0169	144.863	59.161	37.872	2 240.56	90
96	2.599	.3847	.00625	.0163	159.927	61.528	39.973	2 459.42	96
100	2.705	.3697	.00587	.0159	170.481	63.029	41.343	2 605.77	100
104	2.815	.3553	.00551	.0155	181.464	64.471	42.688	2 752.17	104
120	3.300	.3030	.00435	.0143	230.039	69.701	47.835	3 334.11	120
240	10.893	.0918	.00101	.0110	989.254	90.819	75.739	6 878.59	240
360	35.950	.0278	.00029	.0103	3 495.0	97.218	89.699	8 720.43	360
480	118.648	.00843	.00008	.0101	11 764.8	99.157	95.920	9 511.15	480

2% Compound Interest Factors

n	Single Payment — Compound Amount Factor, Find F Given P, F/P	Single Payment — Present Worth Factor, Find P Given F, P/F	Uniform Payment Series — Sinking Fund Factor, Find A Given F, A/F	Uniform Payment Series — Capital Recovery Factor, Find A Given P, A/P	Uniform Payment Series — Compound Amount Factor, Find F Given A, F/A	Uniform Payment Series — Present Worth Factor, Find P Given A, P/A	Arithmetic Gradient — Gradient Uniform Series, Find A Given G, A/G	Arithmetic Gradient — Gradient Present Worth, Find P Given G, P/G	n
1	1.020	.9804	1.0000	1.0200	1.000	0.980	0	0	1
2	1.040	.9612	.4951	.5151	2.020	1.942	0.495	0.961	2
3	1.061	.9423	.3268	.3468	3.060	2.884	0.987	2.846	3
4	1.082	.9238	.2426	.2626	4.122	3.808	1.475	5.617	4
5	1.104	.9057	.1922	.2122	5.204	4.713	1.960	9.240	5
6	1.126	.8880	.1585	.1785	6.308	5.601	2.442	13.679	6
7	1.149	.8706	.1345	.1545	7.434	6.472	2.921	18.903	7
8	1.172	.8535	.1165	.1365	8.583	7.325	3.396	24.877	8
9	1.195	.8368	.1025	.1225	9.755	8.162	3.868	31.571	9
10	1.219	.8203	.0913	.1113	10.950	8.983	4.337	38.954	10
11	1.243	.8043	.0822	.1022	12.169	9.787	4.802	46.996	11
12	1.268	.7885	.0746	.0946	13.412	10.575	5.264	55.669	12
13	1.294	.7730	.0681	.0881	14.680	11.348	5.723	64.946	13
14	1.319	.7579	.0626	.0826	15.974	12.106	6.178	74.798	14
15	1.346	.7430	.0578	.0778	17.293	12.849	6.631	85.200	15
16	1.373	.7284	.0537	.0737	18.639	13.578	7.080	96.127	16
17	1.400	.7142	.0500	.0700	20.012	14.292	7.526	107.553	17
18	1.428	.7002	.0467	.0667	21.412	14.992	7.968	119.456	18
19	1.457	.6864	.0438	.0638	22.840	15.678	8.407	131.812	19
20	1.486	.6730	.0412	.0612	24.297	16.351	8.843	144.598	20
21	1.516	.6598	.0388	.0588	25.783	17.011	9.276	157.793	21
22	1.546	.6468	.0366	.0566	27.299	17.658	9.705	171.377	22
23	1.577	.6342	.0347	.0547	28.845	18.292	10.132	185.328	23
24	1.608	.6217	.0329	.0529	30.422	18.914	10.555	199.628	24
25	1.641	.6095	.0312	.0512	32.030	19.523	10.974	214.256	25
26	1.673	.5976	.0297	.0497	33.671	20.121	11.391	229.196	26
27	1.707	.5859	.0283	.0483	35.344	20.707	11.804	244.428	27
28	1.741	.5744	.0270	.0470	37.051	21.281	12.214	259.936	28
29	1.776	.5631	.0258	.0458	38.792	21.844	12.621	275.703	29
30	1.811	.5521	.0247	.0447	40.568	22.396	13.025	291.713	30
36	2.040	.4902	.0192	.0392	51.994	25.489	15.381	392.036	36
40	2.208	.4529	.0166	.0366	60.402	27.355	16.888	461.989	40
48	2.587	.3865	.0126	.0326	79.353	30.673	19.755	605.961	48
50	2.692	.3715	.0118	.0318	84.579	31.424	20.442	642.355	50
52	2.800	.3571	.0111	.0311	90.016	32.145	21.116	678.779	52
60	3.281	.3048	.00877	.0288	114.051	34.761	23.696	823.692	60
70	4.000	.2500	.00667	.0267	149.977	37.499	26.663	999.829	70
72	4.161	.2403	.00633	.0263	158.056	37.984	27.223	1 034.050	72
80	4.875	.2051	.00516	.0252	193.771	39.744	29.357	1 166.781	80
84	5.277	.1895	.00468	.0247	213.865	40.525	30.361	1 230.413	84
90	5.943	.1683	.00405	.0240	247.155	41.587	31.793	1 322.164	90
96	6.693	.1494	.00351	.0235	284.645	42.529	33.137	1 409.291	96
100	7.245	.1380	.00320	.0232	312.230	43.098	33.986	1 464.747	100
104	7.842	.1275	.00292	.0229	342.090	43.624	34.799	1 518.082	104
120	10.765	.0929	.00205	.0220	488.255	45.355	37.711	1 710.411	120
240	115.887	.00863	.00017	.0202	5 744.4	49.569	47.911	2 374.878	240
360	1 247.5	.00080	.00002	.0200	62 326.8	49.960	49.711	2 483.567	360
480	13 429.8	.00007		.0200	671 442.0	49.996	49.964	2 498.027	480

1½% Compound Interest Factors

n	Single Payment — Compound Amount Factor, Find F Given P, F/P	Single Payment — Present Worth Factor, Find P Given F, P/F	Uniform Payment Series — Sinking Fund Factor, Find A Given F, A/F	Uniform Payment Series — Capital Recovery Factor, Find A Given P, A/P	Uniform Payment Series — Compound Amount Factor, Find F Given A, F/A	Uniform Payment Series — Present Worth Factor, Find P Given A, P/A	Arithmetic Gradient — Gradient Uniform Series, Find A Given G, A/G	Arithmetic Gradient — Gradient Present Worth, Find P Given G, P/G	n
1	1.015	.9852	1.0000	1.0150	1.000	0.985	0	0	1
2	1.030	.9707	.4963	.5113	2.015	1.956	0.496	0.970	2
3	1.046	.9563	.3284	.3434	3.045	2.912	0.990	2.883	3
4	1.061	.9422	.2444	.2594	4.091	3.854	1.481	5.709	4
5	1.077	.9283	.1941	.2091	5.152	4.783	1.970	9.422	5
6	1.093	.9145	.1605	.1755	6.230	5.697	2.456	13.994	6
7	1.110	.9010	.1366	.1516	7.323	6.598	2.940	19.400	7
8	1.126	.8877	.1186	.1336	8.433	7.486	3.422	25.614	8
9	1.143	.8746	.1046	.1196	9.559	8.360	3.901	32.610	9
10	1.161	.8617	.0934	.1084	10.703	9.222	4.377	40.365	10
11	1.178	.8489	.0843	.0993	11.863	10.071	4.851	48.855	11
12	1.196	.8364	.0767	.0917	13.041	10.907	5.322	58.054	12
13	1.214	.8240	.0702	.0852	14.237	11.731	5.791	67.943	13
14	1.232	.8118	.0647	.0797	15.450	12.543	6.258	78.496	14
15	1.250	.7999	.0599	.0749	16.682	13.343	6.722	89.694	15
16	1.269	.7880	.0558	.0708	17.932	14.131	7.184	101.514	16
17	1.288	.7764	.0521	.0671	19.201	14.908	7.643	113.937	17
18	1.307	.7649	.0488	.0638	20.489	15.673	8.100	126.940	18
19	1.327	.7536	.0459	.0609	21.797	16.426	8.554	140.505	19
20	1.347	.7425	.0432	.0582	23.124	17.169	9.005	154.611	20
21	1.367	.7315	.0409	.0559	24.470	17.900	9.455	169.241	21
22	1.388	.7207	.0387	.0537	25.837	18.621	9.902	184.375	22
23	1.408	.7100	.0367	.0517	27.225	19.331	10.346	199.996	23
24	1.430	.6995	.0349	.0499	28.633	20.030	10.788	216.085	24
25	1.451	.6892	.0333	.0483	30.063	20.720	11.227	232.626	25
26	1.473	.6790	.0317	.0467	31.514	21.399	11.664	249.601	26
27	1.495	.6690	.0303	.0453	32.987	22.068	12.099	266.995	27
28	1.517	.6591	.0290	.0440	34.481	22.727	12.531	284.790	28
29	1.540	.6494	.0278	.0428	35.999	23.376	12.961	302.972	29
30	1.563	.6398	.0266	.0416	37.539	24.016	13.388	321.525	30
36	1.709	.5851	.0212	.0362	47.276	27.661	15.901	439.823	36
40	1.814	.5513	.0184	.0334	54.268	29.916	17.528	524.349	40
48	2.043	.4894	.0144	.0294	69.565	34.042	20.666	703.537	48
50	2.105	.4750	.0136	.0286	73.682	35.000	21.428	749.955	50
52	2.169	.4611	.0128	.0278	77.925	35.929	22.179	796.868	52
60	2.443	.4093	.0104	.0254	96.214	39.380	25.093	988.157	60
70	2.835	.3527	.00817	.0232	122.363	43.155	28.529	1 231.15	70
72	2.921	.3423	.00781	.0228	128.076	43.845	29.189	1 279.78	72
80	3.291	.3039	.00655	.0215	152.710	46.407	31.742	1 473.06	80
84	3.493	.2863	.00602	.0210	166.172	47.579	32.967	1 568.50	84
90	3.819	.2619	.00532	.0203	187.929	49.210	34.740	1 709.53	90
96	4.176	.2395	.00472	.0197	211.719	50.702	36.438	1 847.46	96
100	4.432	.2256	.00437	.0194	228.802	51.625	37.529	1 937.43	100
104	4.704	.2126	.00405	.0190	246.932	52.494	38.589	2 025.69	104
120	5.969	.1675	.00302	.0180	331.286	55.498	42.518	2 359.69	120
240	35.632	.0281	.00043	.0154	2 308.8	64.796	59.737	3 870.68	240
360	212.700	.00470	.00007	.0151	14 113.3	66.353	64.966	4 310.71	360
480	1 269.7	.00079	.00001	.0150	84 577.8	66.614	66.288	4 415.74	480

2½%

	Single Payment		Uniform Payment Series				Arithmetic Gradient		
	Compound Amount Factor	Present Worth Factor	Sinking Fund Factor	Capital Recovery Factor	Compound Amount Factor	Present Worth Factor	Gradient Uniform Series	Gradient Present Worth	
n	Find F Given P F/P	Find P Given F P/F	Find A Given F A/F	Find A Given P A/P	Find F Given A F/A	Find P Given A P/A	Find A Given G A/G	Find P Given G P/G	n
1	1.025	.9756	1.0000	1.0250	1.000	0.976	0	0	1
2	1.051	.9518	.4938	.5188	2.025	1.927	0.494	0.952	2
3	1.077	.9286	.3251	.3501	3.076	2.856	0.984	2.809	3
4	1.104	.9060	.2408	.2658	4.153	3.762	1.469	5.527	4
5	1.131	.8839	.1902	.2152	5.256	4.646	1.951	9.062	5
6	1.160	.8623	.1566	.1816	6.388	5.508	2.428	13.374	6
7	1.189	.8413	.1325	.1575	7.547	6.349	2.901	18.421	7
8	1.218	.8207	.1145	.1395	8.736	7.170	3.370	24.166	8
9	1.249	.8007	.1005	.1255	9.955	7.971	3.835	30.572	9
10	1.280	.7812	.0893	.1143	11.203	8.752	4.296	37.603	10
11	1.312	.7621	.0801	.1051	12.483	9.514	4.753	45.224	11
12	1.345	.7436	.0725	.0975	13.796	10.258	5.206	53.403	12
13	1.379	.7254	.0660	.0910	15.140	10.983	5.655	62.108	13
14	1.413	.7077	.0605	.0855	16.519	11.691	6.100	71.309	14
15	1.448	.6905	.0558	.0808	17.932	12.381	6.540	80.975	15
16	1.485	.6736	.0516	.0766	19.380	13.055	6.977	91.080	16
17	1.522	.6572	.0479	.0729	20.865	13.712	7.409	101.595	17
18	1.560	.6412	.0447	.0697	22.386	14.353	7.838	112.495	18
19	1.599	.6255	.0418	.0668	23.946	14.979	8.262	123.754	19
20	1.639	.6103	.0391	.0641	25.545	15.589	8.682	135.349	20
21	1.680	.5954	.0368	.0618	27.183	16.185	9.099	147.257	21
22	1.722	.5809	.0346	.0596	28.863	16.765	9.511	159.455	22
23	1.765	.5667	.0327	.0577	30.584	17.332	9.919	171.922	23
24	1.809	.5529	.0309	.0559	32.349	17.885	10.324	184.638	24
25	1.854	.5394	.0293	.0543	34.158	18.424	10.724	197.584	25
26	1.900	.5262	.0278	.0528	36.012	18.951	11.120	210.740	26
27	1.948	.5134	.0264	.0514	37.912	19.464	11.513	224.088	27
28	1.996	.5009	.0251	.0501	39.860	19.965	11.901	237.612	28
29	2.046	.4887	.0239	.0489	41.856	20.454	12.286	251.294	29
30	2.098	.4767	.0228	.0478	43.903	20.930	12.667	265.120	30
31	2.150	.4651	.0217	.0467	46.000	21.395	13.044	279.073	31
32	2.204	.4538	.0208	.0458	48.150	21.849	13.417	293.140	32
33	2.259	.4427	.0199	.0449	50.354	22.292	13.786	307.306	33
34	2.315	.4319	.0190	.0440	52.613	22.724	14.151	321.559	34
35	2.373	.4214	.0182	.0432	54.928	23.145	14.512	335.886	35
40	2.685	.3724	.0148	.0398	67.402	25.103	16.262	408.221	40
45	3.038	.3292	.0123	.0373	81.516	26.833	17.918	480.806	45
50	3.437	.2909	.0103	.0353	97.484	28.362	19.484	552.607	50
55	3.889	.2572	.00865	.0337	115.551	29.714	20.961	622.827	55
60	4.400	.2273	.00735	.0324	135.991	30.909	22.352	690.865	60
65	4.978	.2009	.00628	.0313	159.118	31.965	23.660	756.280	65
70	5.632	.1776	.00540	.0304	185.284	32.898	24.888	818.763	70
75	6.372	.1569	.00465	.0297	214.888	33.723	26.039	878.114	75
80	7.210	.1387	.00403	.0290	248.382	34.452	27.117	934.217	80
85	8.157	.1226	.00349	.0285	286.278	35.096	28.123	987.026	85
90	9.229	.1084	.00304	.0280	329.154	35.666	29.063	1 036.54	90
95	10.442	.0958	.00265	.0276	377.663	36.169	29.938	1 082.83	95
100	11.814	.0846	.00231	.0273	432.548	36.614	30.752	1 125.97	100

3%

	Single Payment		Uniform Payment Series				Arithmetic Gradient		
	Compound Amount Factor	Present Worth Factor	Sinking Fund Factor	Capital Recovery Factor	Compound Amount Factor	Present Worth Factor	Gradient Uniform Series	Gradient Present Worth	
n	Find F Given P F/P	Find P Given F P/F	Find A Given F A/F	Find A Given P A/P	Find F Given A F/A	Find P Given A P/A	Find A Given G A/G	Find P Given G P/G	n
1	1.030	.9709	1.0300	1.0300	1.000	0.971	0	0	1
2	1.061	.9426	.4926	.5226	2.030	1.913	0.493	0.943	2
3	1.093	.9151	.3235	.3535	3.091	2.829	0.980	2.773	3
4	1.126	.8885	.2390	.2690	4.184	3.717	1.463	5.438	4
5	1.159	.8626	.1884	.2184	5.309	4.580	1.941	8.889	5
6	1.194	.8375	.1546	.1846	6.468	5.417	2.414	13.076	6
7	1.230	.8131	.1305	.1605	7.662	6.230	2.882	17.955	7
8	1.267	.7894	.1125	.1425	8.892	7.020	3.345	23.481	8
9	1.305	.7664	.0984	.1284	10.159	7.786	3.803	29.612	9
10	1.344	.7441	.0872	.1172	11.464	8.530	4.256	36.309	10
11	1.384	.7224	.0781	.1081	12.808	9.253	4.705	43.533	11
12	1.426	.7014	.0705	.1005	14.192	9.954	5.148	51.248	12
13	1.469	.6810	.0640	.0940	15.618	10.635	5.587	59.419	13
14	1.513	.6611	.0585	.0885	17.086	11.296	6.021	68.014	14
15	1.558	.6419	.0538	.0838	18.599	11.938	6.450	77.000	15
16	1.605	.6232	.0496	.0796	20.157	12.561	6.874	86.348	16
17	1.653	.6050	.0460	.0760	21.762	13.166	7.294	96.028	17
18	1.702	.5874	.0427	.0727	23.414	13.754	7.708	106.014	18
19	1.754	.5703	.0398	.0698	25.117	14.324	8.118	116.279	19
20	1.806	.5537	.0372	.0672	26.870	14.877	8.523	126.799	20
21	1.860	.5375	.0349	.0649	28.676	15.415	8.923	137.549	21
22	1.916	.5219	.0327	.0627	30.537	15.937	9.319	148.509	22
23	1.974	.5067	.0308	.0608	32.453	16.444	9.709	159.656	23
24	2.033	.4919	.0290	.0590	34.426	16.936	10.095	170.971	24
25	2.094	.4776	.0274	.0574	36.459	17.413	10.477	182.433	25
26	2.157	.4637	.0259	.0559	38.553	17.877	10.853	194.026	26
27	2.221	.4502	.0246	.0546	40.710	18.327	11.226	205.731	27
28	2.288	.4371	.0233	.0533	42.931	18.764	11.593	217.532	28
29	2.357	.4243	.0221	.0521	45.219	19.188	11.956	229.413	29
30	2.427	.4120	.0210	.0510	47.575	19.600	12.314	241.361	30
31	2.500	.4000	.0200	.0500	50.003	20.000	12.668	253.361	31
32	2.575	.3883	.0190	.0490	52.503	20.389	13.017	265.399	32
33	2.652	.3770	.0182	.0482	55.078	20.766	13.362	277.464	33
34	2.732	.3660	.0173	.0473	57.730	21.132	13.702	289.544	34
35	2.814	.3554	.0165	.0465	60.462	21.487	14.037	301.627	35
40	3.262	.3066	.0133	.0433	75.401	23.115	15.650	361.750	40
45	3.782	.2644	.0108	.0408	92.720	24.519	17.156	420.632	45
50	4.384	.2281	.00887	.0389	112.797	25.730	18.558	477.480	50
55	5.082	.1968	.00735	.0373	136.072	26.774	19.860	531.741	55
60	5.892	.1697	.00613	.0361	163.053	27.676	21.067	583.052	60
65	6.830	.1464	.00515	.0351	194.333	28.453	22.184	631.201	65
70	7.918	.1263	.00434	.0343	230.594	29.123	23.215	676.087	70
75	9.179	.1089	.00367	.0337	272.631	29.702	24.163	717.698	75
80	10.641	.0940	.00311	.0331	321.363	30.201	25.035	756.086	80
85	12.336	.0811	.00265	.0326	377.857	30.631	25.835	791.353	85
90	14.300	.0699	.00226	.0323	443.349	31.002	26.567	823.630	90
95	16.578	.0603	.00193	.0319	519.272	31.323	27.235	853.074	95
100	19.219	.0520	.00165	.0316	607.287	31.599	27.844	879.854	100

3½%

n	Single Payment Compound Amount Factor (Find F Given P) F/P	Single Payment Present Worth Factor (Find P Given F) P/F	Uniform Payment Series Sinking Fund Factor (Find A Given F) A/F	Uniform Payment Series Capital Recovery Factor (Find A Given P) A/P	Uniform Payment Series Compound Amount Factor (Find F Given A) F/A	Uniform Payment Series Present Worth Factor (Find P Given A) P/A	Arithmetic Gradient Uniform Series (Find A Given G) A/G	Arithmetic Gradient Present Worth (Find P Given G) P/G	n
1	1.035	.9662	1.0000	1.0350	1.000	0.966	0	0	1
2	1.071	.9335	.4914	.5264	2.035	1.900	0.491	0.933	2
3	1.109	.9019	.3219	.3569	3.106	2.802	0.977	2.737	3
4	1.148	.8714	.2373	.2723	4.215	3.673	1.457	5.352	4
5	1.188	.8420	.1865	.2215	5.362	4.515	1.931	8.719	5
6	1.229	.8135	.1527	.1877	6.550	5.329	2.400	12.787	6
7	1.272	.7860	.1285	.1635	7.779	6.115	2.862	17.503	7
8	1.317	.7594	.1105	.1455	9.052	6.874	3.320	22.819	8
9	1.363	.7337	.0964	.1314	10.368	7.608	3.771	28.688	9
10	1.411	.7089	.0852	.1202	11.731	8.317	4.217	35.069	10
11	1.460	.6849	.0761	.1111	13.142	9.002	4.657	41.918	11
12	1.511	.6618	.0685	.1035	14.602	9.663	5.091	49.198	12
13	1.564	.6394	.0621	.0971	16.113	10.303	5.520	56.871	13
14	1.619	.6178	.0566	.0916	17.677	10.921	5.943	64.902	14
15	1.675	.5969	.0518	.0868	19.296	11.517	6.361	73.258	15
16	1.734	.5767	.0477	.0827	20.971	12.094	6.773	81.909	16
17	1.795	.5572	.0440	.0790	22.705	12.651	7.179	90.824	17
18	1.857	.5384	.0408	.0758	24.500	13.190	7.580	99.976	18
19	1.922	.5202	.0379	.0729	26.357	13.710	7.975	109.339	19
20	1.990	.5026	.0354	.0704	28.280	14.212	8.365	118.888	20
21	2.059	.4856	.0330	.0680	30.269	14.698	8.749	128.599	21
22	2.132	.4692	.0309	.0659	32.329	15.167	9.128	138.451	22
23	2.206	.4533	.0290	.0640	34.460	15.620	9.502	148.423	23
24	2.283	.4380	.0273	.0623	36.666	16.058	9.870	158.496	24
25	2.363	.4231	.0257	.0607	38.950	16.482	10.233	168.652	25
26	2.446	.4088	.0242	.0592	41.313	16.890	10.590	178.873	26
27	2.532	.3950	.0229	.0579	43.759	17.285	10.942	189.143	27
28	2.620	.3817	.0216	.0566	46.291	17.667	11.289	199.448	28
29	2.712	.3687	.0204	.0554	48.911	18.036	11.631	209.773	29
30	2.807	.3563	.0194	.0544	51.623	18.392	11.967	220.105	30
31	2.905	.3442	.0184	.0534	54.429	18.736	12.299	230.432	31
32	3.007	.3326	.0174	.0524	57.334	19.069	12.625	240.742	32
33	3.112	.3213	.0166	.0516	60.341	19.390	12.946	251.025	33
34	3.221	.3105	.0158	.0508	63.453	19.701	13.262	261.271	34
35	3.334	.3000	.0150	.0500	66.674	20.001	13.573	271.470	35
40	3.959	.2526	.0118	.0468	84.550	21.355	15.055	321.490	40
45	4.702	.2127	.00945	.0445	105.781	22.495	16.417	369.307	45
50	5.585	.1791	.00763	.0426	130.998	23.456	17.666	414.369	50
55	6.633	.1508	.00621	.0412	160.946	24.264	18.808	456.352	55
60	7.878	.1269	.00509	.0401	196.516	24.945	19.848	495.104	60
65	9.357	.1069	.00419	.0392	238.762	25.518	20.793	530.598	65
70	11.113	.0900	.00346	.0385	288.937	26.000	21.650	562.895	70
75	13.199	.0758	.00287	.0379	348.529	26.407	22.423	592.121	75
80	15.676	.0638	.00238	.0374	419.305	26.749	23.120	618.438	80
85	18.618	.0537	.00199	.0370	503.365	27.037	23.747	642.036	85
90	22.112	.0452	.00166	.0367	603.202	27.279	24.308	663.118	90
95	26.262	.0381	.00139	.0364	721.778	27.483	24.811	681.890	95
100	31.191	.0321	.00116	.0362	862.608	27.655	25.259	698.554	100

4%

n	Single Payment Compound Amount Factor (Find F Given P) F/P	Single Payment Present Worth Factor (Find P Given F) P/F	Uniform Payment Series Sinking Fund Factor (Find A Given F) A/F	Uniform Payment Series Capital Recovery Factor (Find A Given P) A/P	Uniform Payment Series Compound Amount Factor (Find F Given A) F/A	Uniform Payment Series Present Worth Factor (Find P Given A) P/A	Arithmetic Gradient Uniform Series (Find A Given G) A/G	Arithmetic Gradient Present Worth (Find P Given G) P/G	n
1	1.040	.9615	1.0400	.0400	1.000	0.962	0	0	1
2	1.082	.9246	.4902	.5302	2.040	1.886	0.490	0.925	2
3	1.125	.8890	.3203	.3603	3.122	2.775	0.974	2.702	3
4	1.170	.8548	.2355	.2755	4.246	3.630	1.451	5.267	4
5	1.217	.8219	.1846	.2246	5.416	4.452	1.922	8.555	5
6	1.265	.7903	.1508	.1908	6.633	5.242	2.386	12.506	6
7	1.316	.7599	.1266	.1666	7.898	6.002	2.843	17.066	7
8	1.369	.7307	.1085	.1485	9.214	6.733	3.294	22.180	8
9	1.423	.7026	.0945	.1345	10.583	7.435	3.739	27.801	9
10	1.480	.6756	.0833	.1233	12.006	8.111	4.177	33.881	10
11	1.539	.6496	.0741	.1141	13.486	8.760	4.609	40.377	11
12	1.601	.6246	.0666	.1066	15.026	9.385	5.034	47.248	12
13	1.665	.6006	.0601	.1001	16.627	9.986	5.453	54.454	13
14	1.732	.5775	.0547	.0947	18.292	10.563	5.866	61.962	14
15	1.801	.5553	.0499	.0899	20.024	11.118	6.272	69.735	15
16	1.873	.5339	.0458	.0858	21.825	11.652	6.672	77.744	16
17	1.948	.5134	.0422	.0822	23.697	12.166	7.066	85.958	17
18	2.026	.4936	.0390	.0790	25.645	12.659	7.453	94.350	18
19	2.107	.4746	.0361	.0761	27.671	13.134	7.834	102.893	19
20	2.191	.4564	.0336	.0736	29.778	13.590	8.209	111.564	20
21	2.279	.4388	.0313	.0713	31.969	14.029	8.578	120.341	21
22	2.370	.4220	.0292	.0692	34.248	14.451	8.941	129.202	22
23	2.465	.4057	.0273	.0673	36.618	14.857	9.297	138.128	23
24	2.563	.3901	.0256	.0656	39.083	15.247	9.648	147.101	24
25	2.666	.3751	.0240	.0640	41.646	15.622	9.993	156.104	25
26	2.772	.3607	.0226	.0626	44.312	15.983	10.331	165.121	26
27	2.883	.3468	.0212	.0612	47.084	16.330	10.664	174.138	27
28	2.999	.3335	.0200	.0600	49.968	16.663	10.991	183.142	28
29	3.119	.3207	.0189	.0589	52.966	16.984	11.312	192.120	29
30	3.243	.3083	.0178	.0578	56.085	17.292	11.627	201.062	30
31	3.373	.2965	.0169	.0569	59.328	17.588	11.937	209.955	31
32	3.508	.2851	.0159	.0559	62.701	17.874	12.241	218.792	32
33	3.648	.2741	.0151	.0551	66.209	18.148	12.540	227.563	33
34	3.794	.2636	.0143	.0543	69.858	18.411	12.832	236.260	34
35	3.946	.2534	.0136	.0536	73.652	18.665	13.120	244.876	35
40	4.801	.2083	.0105	.0505	95.025	19.793	14.476	286.530	40
45	5.841	.1712	.00826	.0483	121.029	20.720	15.705	325.402	45
50	7.107	.1407	.00655	.0466	152.667	21.482	16.812	361.163	50
55	8.646	.1157	.00523	.0452	191.159	22.109	17.807	393.689	55
60	10.520	.0951	.00420	.0442	237.990	22.623	18.697	422.996	60
65	12.799	.0781	.00339	.0434	294.968	23.047	19.491	449.201	65
70	15.572	.0642	.00275	.0427	364.290	23.395	20.196	472.479	70
75	18.945	.0528	.00223	.0422	448.630	23.680	20.821	493.041	75
80	23.050	.0434	.00181	.0418	551.244	23.915	21.372	511.116	80
85	28.044	.0357	.00148	.0415	676.089	24.109	21.857	526.938	85
90	34.119	.0293	.00121	.0412	827.981	24.267	22.283	540.737	90
95	41.511	.0241	.00099	.0410	1 012.8	24.398	22.655	552.730	95
100	50.505	.0198	.00081	.0408	1 237.6	24.505	22.980	563.125	100

331

5%

n	Single Payment Compound Amount Factor F/P	Single Payment Present Worth Factor P/F	Sinking Fund Factor A/F	Capital Recovery Factor A/P	Compound Amount Factor F/A	Present Worth Factor P/A	Gradient Uniform Series A/G	Gradient Present Worth P/G
1	1.050	.9524	1.0000	1.0500	1.000	0.952	0	0
2	1.102	.9070	.4878	.5378	2.050	1.859	0.488	0.907
3	1.158	.8638	.3172	.3672	3.152	2.723	0.967	2.635
4	1.216	.8227	.2320	.2820	4.310	3.546	1.439	5.103
5	1.276	.7835	.1810	.2310	5.526	4.329	1.902	8.237
6	1.340	.7462	.1470	.1970	6.802	5.076	2.358	11.968
7	1.407	.7107	.1228	.1728	8.142	5.786	2.805	16.232
8	1.477	.6768	.1047	.1547	9.549	6.463	3.244	20.970
9	1.551	.6446	.0907	.1407	11.027	7.108	3.676	26.127
10	1.629	.6139	.0795	.1295	12.578	7.722	4.099	31.652
11	1.710	.5847	.0704	.1204	14.207	8.306	4.514	37.499
12	1.796	.5568	.0628	.1128	15.917	8.863	4.922	43.624
13	1.886	.5303	.0565	.1065	17.713	9.394	5.321	49.988
14	1.980	.5051	.0510	.1010	19.599	9.899	5.713	56.553
15	2.079	.4810	.0463	.0963	21.579	10.380	6.097	63.288
16	2.183	.4581	.0423	.0923	23.657	10.838	6.474	70.159
17	2.292	.4363	.0387	.0887	25.840	11.274	6.842	77.140
18	2.407	.4155	.0355	.0855	28.132	11.690	7.203	84.204
19	2.527	.3957	.0327	.0827	30.539	12.085	7.557	91.327
20	2.653	.3769	.0302	.0802	33.066	12.462	7.903	98.488
21	2.786	.3589	.0280	.0780	35.719	12.821	8.242	105.667
22	2.925	.3419	.0260	.0760	38.505	13.163	8.573	112.846
23	3.072	.3256	.0241	.0741	41.430	13.489	8.897	120.008
24	3.225	.3101	.0225	.0725	44.502	13.799	9.214	127.140
25	3.386	.2953	.0210	.0710	47.727	14.094	9.524	134.227
26	3.556	.2812	.0196	.0696	51.113	14.375	9.827	141.258
27	3.733	.2678	.0183	.0683	54.669	14.643	10.122	148.222
28	3.920	.2551	.0171	.0671	58.402	14.898	10.411	155.110
29	4.116	.2429	.0160	.0660	62.323	15.141	10.694	161.912
30	4.322	.2314	.0151	.0651	66.439	15.372	10.969	168.622
31	4.538	.2204	.0141	.0641	70.761	15.593	11.238	175.233
32	4.765	.2099	.0133	.0633	75.299	15.803	11.501	181.739
33	5.003	.1999	.0125	.0625	80.063	16.003	11.757	188.135
34	5.253	.1904	.0118	.0618	85.067	16.193	12.006	194.416
35	5.516	.1813	.0111	.0611	90.320	16.374	12.250	200.580
40	7.040	.1420	.00828	.0583	120.799	17.159	13.377	229.545
45	8.985	.1113	.00626	.0563	159.699	17.774	14.364	255.314
50	11.467	.0872	.00478	.0548	209.347	18.256	15.223	277.914
55	14.636	.0683	.00367	.0537	272.711	18.633	15.966	297.510
60	18.679	.0535	.00283	.0528	353.582	18.929	16.606	314.343
65	23.840	.0419	.00219	.0522	456.795	19.161	17.154	328.691
70	30.426	.0329	.00170	.0517	588.525	19.343	17.621	340.841
75	38.832	.0258	.00132	.0513	756.649	19.485	18.018	351.072
80	49.561	.0202	.00103	.0510	971.222	19.596	18.353	359.646
85	63.254	.0158	.00080	.0508	1 245.1	19.684	18.635	366.800
90	80.730	.0124	.00063	.0506	1 594.6	19.752	18.871	372.749
95	103.034	.00971	.00049	.0505	2 040.7	19.806	19.069	377.677
100	131.500	.00760	.00038	.0504	2 610.0	19.848	19.234	381.749

4½%

n	Single Payment Compound Amount Factor F/P	Single Payment Present Worth Factor P/F	Sinking Fund Factor A/F	Capital Recovery Factor A/P	Compound Amount Factor F/A	Present Worth Factor P/A	Gradient Uniform Series A/G	Gradient Present Worth P/G
1	1.045	.9569	1.0000	1.0450	1.000	0.957	0	0
2	1.092	.9157	.4890	.5340	2.045	1.873	0.489	0.916
3	1.141	.8763	.3188	.3638	3.137	2.749	0.971	2.668
4	1.193	.8386	.2337	.2787	4.278	3.588	1.445	5.184
5	1.246	.8025	.1828	.2278	5.471	4.390	1.912	8.394
6	1.302	.7679	.1489	.1939	6.717	5.158	2.372	12.233
7	1.361	.7348	.1247	.1697	8.019	5.893	2.824	16.642
8	1.422	.7032	.1066	.1516	9.380	6.596	3.269	21.564
9	1.486	.6729	.0926	.1376	10.802	7.269	3.707	26.948
10	1.553	.6439	.0814	.1264	12.288	7.913	4.138	32.743
11	1.623	.6162	.0722	.1172	13.841	8.529	4.562	38.905
12	1.696	.5897	.0647	.1097	15.464	9.119	4.978	45.391
13	1.772	.5643	.0583	.1033	17.160	9.683	5.387	52.163
14	1.852	.5400	.0528	.0978	18.932	10.223	5.789	59.182
15	1.935	.5167	.0481	.0931	20.784	10.740	6.184	66.416
16	2.022	.4945	.0440	.0890	22.719	11.234	6.572	73.833
17	2.113	.4732	.0404	.0854	24.742	11.707	6.953	81.404
18	2.208	.4528	.0372	.0822	26.855	12.160	7.327	89.102
19	2.308	.4333	.0344	.0794	29.064	12.593	7.695	96.901
20	2.412	.4146	.0319	.0769	31.371	13.008	8.055	104.779
21	2.520	.3968	.0296	.0746	33.783	13.405	8.409	112.715
22	2.634	.3797	.0275	.0725	36.303	13.784	8.755	120.689
23	2.752	.3634	.0257	.0707	38.937	14.148	9.096	128.682
24	2.876	.3477	.0240	.0690	41.689	14.495	9.429	136.680
25	3.005	.3327	.0224	.0674	44.565	14.828	9.756	144.665
26	3.141	.3184	.0210	.0660	47.571	15.147	10.077	152.625
27	3.282	.3047	.0197	.0647	50.711	15.451	10.391	160.547
28	3.430	.2916	.0185	.0635	53.993	15.743	10.698	168.420
29	3.584	.2790	.0174	.0624	57.423	16.022	10.999	176.232
30	3.745	.2670	.0164	.0614	61.007	16.289	11.295	183.975
31	3.914	.2555	.0154	.0604	64.752	16.544	11.583	191.640
32	4.090	.2445	.0146	.0596	68.666	16.789	11.866	199.220
33	4.274	.2340	.0137	.0587	72.756	17.023	12.143	206.707
34	4.466	.2239	.0130	.0580	77.030	17.247	12.414	214.095
35	4.667	.2143	.0123	.0573	81.497	17.461	12.679	221.380
40	5.816	.1719	.00934	.0543	107.030	18.402	13.917	256.098
45	7.248	.1380	.00720	.0522	138.850	19.156	15.020	287.732
50	9.033	.1107	.00560	.0506	178.503	19.762	15.998	316.145
55	11.256	.0888	.00439	.0494	227.918	20.248	16.860	341.375
60	14.027	.0713	.00345	.0485	289.497	20.638	17.617	363.571
65	17.481	.0572	.00273	.0477	366.237	20.951	18.278	382.946
70	21.784	.0459	.00217	.0472	461.869	21.202	18.854	399.750
75	27.147	.0368	.00172	.0467	581.043	21.404	19.354	414.242
80	33.830	.0296	.00137	.0464	729.556	21.565	19.785	426.680
85	42.158	.0237	.00109	.0461	914.630	21.695	20.157	437.309
90	52.537	.0190	.00087	.0459	1 145.3	21.799	20.476	446.359
95	65.471	.0153	.00070	.0457	1 432.7	21.883	20.749	454.039
100	81.588	.0123	.00056	.0456	1 790.9	21.950	20.981	460.537

Compound Interest Factors — 7%

	Single Payment		Uniform Payment Series				Arithmetic Gradient		
	Compound Amount Factor	Present Worth Factor	Sinking Fund Factor	Capital Recovery Factor	Compound Amount Factor	Present Worth Factor	Gradient Uniform Series	Gradient Present Worth	
n	Find F Given P F/P	Find P Given F P/F	Find A Given F A/F	Find A Given P A/P	Find F Given A F/A	Find P Given A P/A	Find A Given G A/G	Find P Given G P/G	n
1	1.070	.9346	1.0000	1.0700	1.000	0.935	0	0	1
2	1.145	.8734	.4831	.5531	2.070	1.808	0.483	0.873	2
3	1.225	.8163	.3111	.3811	3.215	2.624	0.955	2.506	3
4	1.311	.7629	.2252	.2952	4.440	3.387	1.416	4.795	4
5	1.403	.7130	.1739	.2439	5.751	4.100	1.865	7.647	5
6	1.501	.6663	.1398	.2098	7.153	4.767	2.303	10.978	6
7	1.606	.6227	.1156	.1856	8.654	5.389	2.730	14.715	7
8	1.718	.5820	.0975	.1675	10.260	5.971	3.147	18.789	8
9	1.838	.5439	.0835	.1535	11.978	6.515	3.552	23.140	9
10	1.967	.5083	.0724	.1424	13.816	7.024	3.946	27.716	10
11	2.105	.4751	.0634	.1334	15.784	7.499	4.330	32.467	11
12	2.252	.4440	.0559	.1259	17.888	7.943	4.703	37.351	12
13	2.410	.4150	.0497	.1197	20.141	8.358	5.065	42.330	13
14	2.579	.3878	.0443	.1143	22.551	8.745	5.417	47.372	14
15	2.759	.3624	.0398	.1098	25.129	9.108	5.758	52.446	15
16	2.952	.3387	.0359	.1059	27.888	9.447	6.090	57.527	16
17	3.159	.3166	.0324	.1024	30.840	9.763	6.411	62.592	17
18	3.380	.2959	.0294	.0994	33.999	10.059	6.722	67.622	18
19	3.617	.2765	.0268	.0968	37.379	10.336	7.024	72.599	19
20	3.870	.2584	.0244	.0944	40.996	10.594	7.316	77.509	20
21	4.141	.2415	.0223	.0923	44.865	10.836	7.599	82.339	21
22	4.430	.2257	.0204	.0904	49.006	11.061	7.872	87.079	22
23	4.741	.2109	.0187	.0887	53.436	11.272	8.137	91.720	23
24	5.072	.1971	.0172	.0872	58.177	11.469	8.392	96.255	24
25	5.427	.1842	.0158	.0858	63.249	11.654	8.639	100.677	25
26	5.807	.1722	.0146	.0846	68.677	11.826	8.877	104.981	26
27	6.214	.1609	.0134	.0834	74.484	11.987	9.107	109.166	27
28	6.649	.1504	.0124	.0824	80.698	12.137	9.329	113.227	28
29	7.114	.1406	.0114	.0814	87.347	12.278	9.543	117.162	29
30	7.612	.1314	.0106	.0806	94.461	12.409	9.749	120.972	30
31	8.145	.1228	.00980	.0798	102.073	12.532	9.947	124.655	31
32	8.715	.1147	.00907	.0791	110.218	12.647	10.138	128.212	32
33	9.325	.1072	.00841	.0784	118.934	12.754	10.322	131.644	33
34	9.978	.1002	.00780	.0778	128.259	12.854	10.499	134.951	34
35	10.677	.0937	.00723	.0772	138.237	12.948	10.669	138.135	35
40	14.974	.0668	.00501	.0750	199.636	13.332	11.423	152.293	40
45	21.002	.0476	.00350	.0735	285.750	13.606	12.036	163.756	45
50	29.457	.0339	.00246	.0725	406.530	13.801	12.529	172.905	50
55	41.315	.0242	.00174	.0717	575.930	13.940	12.921	180.124	55
60	57.947	.0173	.00123	.0712	813.523	14.039	13.232	185.768	60
65	81.273	.0123	.00087	.0709	1 146.8	14.110	13.476	190.145	65
70	113.990	.00877	.00062	.0706	1 614.1	14.160	13.666	193.519	70
75	159.877	.00625	.00044	.0704	2 269.7	14.196	13.814	196.104	75
80	224.235	.00446	.00031	.0703	3 189.1	14.222	13.927	198.075	80
85	314.502	.00318	.00022	.0702	4 478.6	14.240	14.015	199.572	85
90	441.105	.00227	.00016	.0702	6 287.2	14.253	14.081	200.704	90
95	618.673	.00162	.00011	.0701	8 823.9	14.263	14.132	201.558	95
100	867.720	.00115	.00008	.0701	12 381.7	14.269	14.170	202.200	100

Compound Interest Factors — 6%

	Single Payment		Uniform Payment Series				Arithmetic Gradient		
	Compound Amount Factor	Present Worth Factor	Sinking Fund Factor	Capital Recovery Factor	Compound Amount Factor	Present Worth Factor	Gradient Uniform Series	Gradient Present Worth	
n	Find F Given P F/P	Find P Given F P/F	Find A Given F A/F	Find A Given P A/P	Find F Given A F/A	Find P Given A P/A	Find A Given G A/G	Find P Given G P/G	n
1	1.060	.9434	1.0000	1.0600	1.000	0.943	0	0	1
2	1.124	.8900	.4854	.5454	2.060	1.833	0.485	0.890	2
3	1.191	.8396	.3141	.3741	3.184	2.673	0.961	2.569	3
4	1.262	.7921	.2286	.2886	4.375	3.465	1.427	4.945	4
5	1.338	.7473	.1774	.2374	5.637	4.212	1.884	7.934	5
6	1.419	.7050	.1434	.2034	6.975	4.917	2.330	11.459	6
7	1.504	.6651	.1191	.1791	8.394	5.582	2.768	15.450	7
8	1.594	.6274	.1010	.1610	9.897	6.210	3.195	19.841	8
9	1.689	.5919	.0870	.1470	11.491	6.802	3.613	24.577	9
10	1.791	.5584	.0759	.1359	13.181	7.360	4.022	29.602	10
11	1.898	.5268	.0668	.1268	14.972	7.887	4.421	34.870	11
12	2.012	.4970	.0593	.1193	16.870	8.384	4.811	40.337	12
13	2.133	.4688	.0530	.1130	18.882	8.853	5.192	45.963	13
14	2.261	.4423	.0476	.1076	21.015	9.295	5.564	51.713	14
15	2.397	.4173	.0430	.1030	23.276	9.712	5.926	57.554	15
16	2.540	.3936	.0390	.0990	25.672	10.106	6.279	63.459	16
17	2.693	.3714	.0354	.0954	28.213	10.477	6.624	69.401	17
18	2.854	.3503	.0324	.0924	30.906	10.828	6.960	75.357	18
19	3.026	.3305	.0296	.0896	33.760	11.158	7.287	81.306	19
20	3.207	.3118	.0272	.0872	36.786	11.470	7.605	87.230	20
21	3.400	.2942	.0250	.0850	39.993	11.764	7.915	93.113	21
22	3.604	.2775	.0230	.0830	43.392	12.042	8.217	98.941	22
23	3.820	.2618	.0213	.0813	46.996	12.303	8.510	104.700	23
24	4.049	.2470	.0197	.0797	50.815	12.550	8.795	110.381	24
25	4.292	.2330	.0182	.0782	54.864	12.783	9.072	115.973	25
26	4.549	.2198	.0169	.0769	59.156	13.003	9.341	121.468	26
27	4.822	.2074	.0157	.0757	63.706	13.211	9.603	126.860	27
28	5.112	.1956	.0146	.0746	68.528	13.406	9.857	132.142	28
29	5.418	.1846	.0136	.0736	73.640	13.591	10.103	137.309	29
30	5.743	.1741	.0126	.0726	79.058	13.765	10.342	142.359	30
31	6.088	.1643	.0118	.0718	84.801	13.929	10.574	147.286	31
32	6.453	.1550	.0110	.0710	90.890	14.084	10.799	152.090	32
33	6.841	.1462	.0103	.0703	97.343	14.230	11.017	156.768	33
34	7.251	.1379	.00960	.0696	104.184	14.368	11.228	161.319	34
35	7.686	.1301	.00897	.0690	111.435	14.498	11.432	165.743	35
40	10.286	.0972	.00646	.0665	154.762	15.046	12.359	185.957	40
45	13.765	.0727	.00470	.0647	212.743	15.456	13.141	203.109	45
50	18.420	.0543	.00344	.0634	290.335	15.762	13.796	217.457	50
55	24.650	.0406	.00254	.0625	394.171	15.991	14.341	229.322	55
60	32.988	.0303	.00188	.0619	533.126	16.161	14.791	239.043	60
65	44.145	.0227	.00139	.0614	719.080	16.289	15.160	246.945	65
70	59.076	.0169	.00103	.0610	967.928	16.385	15.461	253.327	70
75	79.057	.0126	.00077	.0608	1 300.9	16.456	15.706	258.453	75
80	105.796	.00945	.00057	.0606	1 746.6	16.509	15.903	262.549	80
85	141.578	.00706	.00043	.0604	2 343.0	16.549	16.062	265.810	85
90	189.464	.00528	.00032	.0603	3 141.1	16.579	16.189	268.395	90
95	253.545	.00394	.00024	.0602	4 209.1	16.601	16.290	270.437	95
100	339.300	.00295	.00018	.0602	5 638.3	16.618	16.371	272.047	100

8% — Compound Interest Factors

n	Single Payment F/P (F given P)	Single Payment P/F (P given F)	Sinking Fund A/F (A given F)	Capital Recovery A/P (A given P)	Compound Amount F/A (F given A)	Present Worth P/A (P given A)	Gradient Uniform A/G (A given G)	Gradient Present Worth P/G (P given G)	n
1	1.080	.9259	1.0000	1.0800	1.000	0.926	0	0	1
2	1.166	.8573	.4808	.5608	2.080	1.783	0.481	0.857	2
3	1.260	.7938	.3080	.3880	3.246	2.577	0.949	2.445	3
4	1.360	.7350	.2219	.3019	4.506	3.312	1.404	4.650	4
5	1.469	.6806	.1705	.2505	5.867	3.993	1.846	7.372	5
6	1.587	.6302	.1363	.2163	7.336	4.623	2.276	10.523	6
7	1.714	.5835	.1121	.1921	8.923	5.206	2.694	14.024	7
8	1.851	.5403	.0940	.1740	10.637	5.747	3.099	17.806	8
9	1.999	.5002	.0801	.1601	12.488	6.247	3.491	21.808	9
10	2.159	.4632	.0690	.1490	14.487	6.710	3.871	25.977	10
11	2.332	.4289	.0601	.1401	16.645	7.139	4.240	30.266	11
12	2.518	.3971	.0527	.1327	18.977	7.536	4.596	34.634	12
13	2.720	.3677	.0465	.1265	21.495	7.904	4.940	39.046	13
14	2.937	.3405	.0413	.1213	24.215	8.244	5.273	43.472	14
15	3.172	.3152	.0368	.1168	27.152	8.559	5.594	47.886	15
16	3.426	.2919	.0330	.1130	30.324	8.851	5.905	52.264	16
17	3.700	.2703	.0296	.1096	33.750	9.122	6.204	56.588	17
18	3.996	.2502	.0267	.1067	37.450	9.372	6.492	60.843	18
19	4.316	.2317	.0241	.1041	41.446	9.604	6.770	65.013	19
20	4.661	.2145	.0219	.1019	45.762	9.818	7.037	69.090	20
21	5.034	.1987	.0198	.0998	50.423	10.017	7.294	73.063	21
22	5.437	.1839	.0180	.0980	55.457	10.201	7.541	76.926	22
23	5.871	.1703	.0164	.0964	60.893	10.371	7.779	80.673	23
24	6.341	.1577	.0150	.0950	66.765	10.529	8.007	84.300	24
25	6.848	.1460	.0137	.0937	73.106	10.675	8.225	87.804	25
26	7.396	.1352	.0125	.0925	79.954	10.810	8.435	91.184	26
27	7.988	.1252	.0114	.0914	87.351	10.935	8.636	94.439	27
28	8.627	.1159	.0105	.0905	95.339	11.051	8.829	97.569	28
29	9.317	.1073	.00962	.0896	103.966	11.158	9.013	100.574	29
30	10.063	.0994	.00883	.0888	113.283	11.258	9.190	103.456	30
31	10.868	.0920	.00811	.0881	123.346	11.350	9.358	106.216	31
32	11.737	.0852	.00745	.0875	134.214	11.435	9.520	108.858	32
33	12.676	.0789	.00685	.0869	145.951	11.514	9.674	111.382	33
34	13.690	.0730	.00630	.0863	158.627	11.587	9.821	113.792	34
35	14.785	.0676	.00580	.0858	172.317	11.655	9.961	116.092	35
40	21.725	.0460	.00386	.0839	259.057	11.925	10.570	126.042	40
45	31.920	.0313	.00259	.0826	386.506	12.108	11.045	133.733	45
50	46.902	.0213	.00174	.0817	573.771	12.233	11.411	139.593	50
55	68.914	.0145	.00118	.0812	848.925	12.319	11.690	144.006	55
60	101.257	.00988	.00080	.0808	1 253.2	12.377	11.902	147.300	60
65	148.780	.00672	.00054	.0805	1 847.3	12.416	12.060	149.739	65
70	218.607	.00457	.00037	.0804	2 720.1	12.443	12.178	151.533	70
75	321.205	.00311	.00025	.0802	4 002.6	12.461	12.266	152.845	75
80	471.956	.00212	.00017	.0802	5 887.0	12.474	12.330	153.800	80
85	693.458	.00144	.00012	.0801	8 655.7	12.482	12.377	154.492	85
90	1 018.9	.00098	.00008	.0801	12 724.0	12.488	12.412	154.993	90
95	1 497.1	.00067	.00005	.0801	18 701.6	12.492	12.437	155.352	95
100	2 199.8	.00045	.00004	.0800	27 484.6	12.494	12.455	155.611	100

9% — Compound Interest Factors

n	Single Payment F/P (F given P)	Single Payment P/F (P given F)	Sinking Fund A/F (A given F)	Capital Recovery A/P (A given P)	Compound Amount F/A (F given A)	Present Worth P/A (P given A)	Gradient Uniform A/G (A given G)	Gradient Present Worth P/G (P given G)	n
1	1.090	.9174	1.0000	1.0900	1.000	0.917	0	0	1
2	1.188	.8417	.4785	.5685	2.090	1.759	0.478	0.842	2
3	1.295	.7722	.3051	.3951	3.278	2.531	0.943	2.386	3
4	1.412	.7084	.2187	.3087	4.573	3.240	1.393	4.511	4
5	1.539	.6499	.1671	.2571	5.985	3.890	1.828	7.111	5
6	1.677	.5963	.1329	.2229	7.523	4.486	2.250	10.092	6
7	1.828	.5470	.1087	.1987	9.200	5.033	2.657	13.375	7
8	1.993	.5019	.0907	.1807	11.028	5.535	3.051	16.888	8
9	2.172	.4604	.0768	.1668	13.021	5.995	3.431	20.571	9
10	2.367	.4224	.0658	.1558	15.193	6.418	3.798	24.373	10
11	2.580	.3875	.0569	.1469	17.560	6.805	4.151	28.248	11
12	2.813	.3555	.0497	.1397	20.141	7.161	4.491	32.159	12
13	3.066	.3262	.0436	.1336	22.953	7.487	4.818	36.073	13
14	3.342	.2992	.0384	.1284	26.019	7.786	5.133	39.963	14
15	3.642	.2745	.0341	.1241	29.361	8.061	5.435	43.807	15
16	3.970	.2519	.0303	.1203	33.003	8.313	5.724	47.585	16
17	4.328	.2311	.0270	.1170	36.974	8.544	6.002	51.282	17
18	4.717	.2120	.0242	.1142	41.301	8.756	6.269	54.886	18
19	5.142	.1945	.0217	.1117	46.019	8.950	6.524	58.387	19
20	5.604	.1784	.0195	.1095	51.160	9.129	6.767	61.777	20
21	6.109	.1637	.0176	.1076	56.765	9.292	7.001	65.051	21
22	6.659	.1502	.0159	.1059	62.873	9.442	7.223	68.205	22
23	7.258	.1378	.0144	.1044	69.532	9.580	7.436	71.236	23
24	7.911	.1264	.0130	.1030	76.790	9.707	7.638	74.143	24
25	8.623	.1160	.0118	.1018	84.701	9.823	7.832	76.927	25
26	9.399	.1064	.0107	.1007	93.324	9.929	8.016	79.586	26
27	10.245	.0976	.00973	.0997	102.723	10.027	8.191	82.124	27
28	11.167	.0895	.00885	.0989	112.968	10.116	8.357	84.542	28
29	12.172	.0822	.00806	.0981	124.136	10.198	8.515	86.842	29
30	13.268	.0754	.00734	.0973	136.308	10.274	8.666	89.028	30
31	14.462	.0691	.00669	.0967	149.575	10.343	8.808	91.102	31
32	15.763	.0634	.00610	.0961	164.037	10.406	8.944	93.069	32
33	17.182	.0582	.00556	.0956	179.801	10.464	9.072	94.931	33
34	18.728	.0534	.00508	.0951	196.983	10.518	9.193	96.693	34
35	20.414	.0490	.00464	.0946	215.711	10.567	9.308	98.359	35
40	31.409	.0318	.00296	.0930	337.883	10.757	9.796	105.376	40
45	48.327	.0207	.00190	.0919	525.860	10.881	10.160	110.556	45
50	74.358	.0134	.00123	.0912	815.085	10.962	10.430	114.325	50
55	114.409	.00874	.00079	.0908	1 260.1	11.014	10.626	117.036	55
60	176.032	.00568	.00051	.0905	1 944.8	11.048	10.768	118.968	60
65	270.847	.00369	.00033	.0903	2 998.3	11.070	10.870	120.334	65
70	416.731	.00240	.00022	.0902	4 619.2	11.084	10.943	121.294	70
75	641.193	.00156	.00014	.0901	7 113.3	11.094	10.994	121.965	75
80	986.555	.00101	.00009	.0901	10 950.6	11.100	11.030	122.431	80
85	1 517.9	.00066	.00006	.0901	16 854.9	11.104	11.055	122.753	85
90	2 335.5	.00043	.00004	.0900	25 939.3	11.106	11.073	122.976	90
95	3 593.5	.00028	.00003	.0900	39 916.8	11.108	11.085	123.129	95
100	5 529.1	.00018	.00002	.0900	61 422.9	11.109	11.093	123.233	100

Compound Interest Factors — **10%**

	Single Payment		Uniform Payment Series				Arithmetic Gradient		
	Compound Amount Factor	Present Worth Factor	Sinking Fund Factor	Capital Recovery Factor	Compound Amount Factor	Present Worth Factor	Gradient Uniform Series	Gradient Present Worth	
n	Find F Given P F/P	Find P Given F P/F	Find A Given F A/F	Find A Given P A/P	Find F Given A F/A	Find P Given A P/A	Find A Given G A/G	Find P Given G P/G	n
1	1.100	.9091	1.0000	1.1000	1.000	0.909	0	0	1
2	1.210	.8264	.4762	.5762	2.100	1.736	0.476	0.826	2
3	1.331	.7513	.3021	.4021	3.310	2.487	0.937	2.329	3
4	1.464	.6830	.2155	.3155	4.641	3.170	1.381	4.378	4
5	1.611	.6209	.1638	.2638	6.105	3.791	1.810	6.862	5
6	1.772	.5645	.1296	.2296	7.716	4.355	2.224	9.684	6
7	1.949	.5132	.1054	.2054	9.487	4.868	2.622	12.763	7
8	2.144	.4665	.0874	.1874	11.436	5.335	3.004	16.029	8
9	2.358	.4241	.0736	.1736	13.579	5.759	3.372	19.421	9
10	2.594	.3855	.0627	.1627	15.937	6.145	3.725	22.891	10
11	2.853	.3505	.0540	.1540	18.531	6.495	4.064	26.396	11
12	3.138	.3186	.0468	.1468	21.384	6.814	4.388	29.901	12
13	3.452	.2897	.0408	.1408	24.523	7.103	4.699	33.377	13
14	3.797	.2633	.0357	.1357	27.975	7.367	4.996	36.801	14
15	4.177	.2394	.0315	.1315	31.772	7.606	5.279	40.152	15
16	4.595	.2176	.0278	.1278	35.950	7.824	5.549	43.416	16
17	5.054	.1978	.0247	.1247	40.545	8.022	5.807	46.582	17
18	5.560	.1799	.0219	.1219	45.599	8.201	6.053	49.640	18
19	6.116	.1635	.0195	.1195	51.159	8.365	6.286	52.583	19
20	6.728	.1486	.0175	.1175	57.275	8.514	6.508	55.407	20
21	7.400	.1351	.0156	.1156	64.003	8.649	6.719	58.110	21
22	8.140	.1228	.0140	.1140	71.403	8.772	6.919	60.689	22
23	8.954	.1117	.0126	.1126	79.543	8.883	7.108	63.146	23
24	9.850	.1015	.0113	.1113	88.497	8.985	7.288	65.481	24
25	10.835	.0923	.0102	.1102	98.347	9.077	7.458	67.696	25
26	11.918	.0839	.00916	.1092	109.182	9.161	7.619	69.794	26
27	13.110	.0763	.00826	.1083	121.100	9.237	7.770	71.777	27
28	14.421	.0693	.00745	.1075	134.210	9.307	7.914	73.650	28
29	15.863	.0630	.00673	.1067	148.631	9.370	8.049	75.415	29
30	17.449	.0573	.00608	.1061	164.494	9.427	8.176	77.077	30
31	19.194	.0521	.00550	.1055	181.944	9.479	8.296	78.640	31
32	21.114	.0474	.00497	.1050	201.138	9.526	8.409	80.108	32
33	23.225	.0431	.00450	.1045	222.252	9.569	8.515	81.486	33
34	25.548	.0391	.00407	.1041	245.477	9.609	8.615	82.777	34
35	28.102	.0356	.00369	.1037	271.025	9.644	8.709	83.987	35
40	45.259	.0221	.00226	.1023	442.593	9.779	9.096	88.953	40
45	72.891	.0137	.00139	.1014	718.905	9.863	9.374	92.454	45
50	117.391	.00852	.00086	.1009	1 163.9	9.915	9.570	94.889	50
55	189.059	.00529	.00053	.1005	1 880.6	9.947	9.708	96.562	55
60	304.482	.00328	.00033	.1003	3 034.8	9.967	9.802	97.701	60
65	490.371	.00204	.00020	.1002	4 893.7	9.980	9.867	98.471	65
70	789.748	.00127	.00013	.1001	7 887.5	9.987	9.911	98.987	70
75	1 271.9	.00079	.00008	.1001	12 709.0	9.992	9.941	99.332	75
80	2 048.4	.00049	.00005	.1000	20 474.0	9.995	9.961	99.561	80
85	3 299.0	.00030	.00003	.1000	32 979.7	9.997	9.974	99.712	85
90	5 313.0	.00019	.00002	.1000	53 120.3	9.998	9.983	99.812	90
95	8 556.7	.00012	.00001	.1000	85 556.9	9.999	9.989	99.877	95
100	13 780.6	.00007	.00001	.1000	137 796.3	9.999	9.993	99.920	100

Compound Interest Factors — **12%**

	Single Payment		Uniform Payment Series				Arithmetic Gradient		
	Compound Amount Factor	Present Worth Factor	Sinking Fund Factor	Capital Recovery Factor	Compound Amount Factor	Present Worth Factor	Gradient Uniform Series	Gradient Present Worth	
n	Find F Given P F/P	Find P Given F P/F	Find A Given F A/F	Find A Given P A/P	Find F Given A F/A	Find P Given A P/A	Find A Given G A/G	Find P Given G P/G	n
1	1.120	.8929	1.0000	1.1200	1.000	0.893	0	0	1
2	1.254	.7972	.4717	.5917	2.120	1.690	0.472	0.797	2
3	1.405	.7118	.2963	.4163	3.374	2.402	0.925	2.221	3
4	1.574	.6355	.2092	.3292	4.779	3.037	1.359	4.127	4
5	1.762	.5674	.1574	.2774	6.353	3.605	1.775	6.397	5
6	1.974	.5066	.1232	.2432	8.115	4.111	2.172	8.930	6
7	2.211	.4523	.0991	.2191	10.089	4.564	2.551	11.644	7
8	2.476	.4039	.0813	.2013	12.300	4.968	2.913	14.471	8
9	2.773	.3606	.0677	.1877	14.776	5.328	3.257	17.356	9
10	3.106	.3220	.0570	.1770	17.549	5.650	3.585	20.254	10
11	3.479	.2875	.0484	.1684	20.655	5.938	3.895	23.129	11
12	3.896	.2567	.0414	.1614	24.133	6.194	4.190	25.952	12
13	4.363	.2292	.0357	.1557	28.029	6.424	4.468	28.702	13
14	4.887	.2046	.0309	.1509	32.393	6.628	4.732	31.362	14
15	5.474	.1827	.0268	.1468	37.280	6.811	4.980	33.920	15
16	6.130	.1631	.0234	.1434	42.753	6.974	5.215	36.367	16
17	6.866	.1456	.0205	.1405	48.884	7.120	5.435	38.697	17
18	7.690	.1300	.0179	.1379	55.750	7.250	5.643	40.908	18
19	8.613	.1161	.0158	.1358	63.440	7.366	5.838	42.998	19
20	9.646	.1037	.0139	.1339	72.052	7.469	6.020	44.968	20
21	10.804	.0926	.0122	.1322	81.699	7.562	6.191	46.819	21
22	12.100	.0826	.0108	.1308	92.503	7.645	6.351	48.554	22
23	13.552	.0738	.00956	.1296	104.603	7.718	6.501	50.178	23
24	15.179	.0659	.00846	.1285	118.155	7.784	6.641	51.693	24
25	17.000	.0588	.00750	.1275	133.334	7.843	6.771	53.105	25
26	19.040	.0525	.00665	.1267	150.334	7.896	6.892	54.418	26
27	21.325	.0469	.00590	.1259	169.374	7.943	7.005	55.637	27
28	23.884	.0419	.00524	.1252	190.699	7.984	7.110	56.767	28
29	26.750	.0374	.00466	.1247	214.583	8.022	7.207	57.814	29
30	29.960	.0334	.00414	.1241	241.333	8.055	7.297	58.782	30
31	33.555	.0298	.00369	.1237	271.293	8.085	7.381	59.676	31
32	37.582	.0266	.00328	.1233	304.848	8.112	7.459	60.501	32
33	42.092	.0238	.00292	.1229	342.429	8.135	7.530	61.261	33
34	47.143	.0212	.00260	.1226	384.521	8.157	7.596	61.961	34
35	52.800	.0189	.00232	.1223	431.663	8.176	7.658	62.605	35
40	93.051	.0107	.00130	.1213	767.091	8.244	7.899	65.116	40
45	163.988	.00610	.00074	.1207	1 358.2	8.283	8.057	66.734	45
50	289.002	.00346	.00042	.1204	2 400.0	8.304	8.160	67.762	50
55	509.321	.00196	.00024	.1202	4 236.0	8.317	8.225	68.408	55
60	897.597	.00111	.00013	.1201	7 471.6	8.324	8.266	68.810	60
65	1 581.9	.00063	.00008	.1201	13 173.9	8.328	8.292	69.058	65
70	2 787.8	.00036	.00004	.1200	23 223.3	8.330	8.308	69.210	70
75	4 913.1	.00020	.00002	.1200	40 933.8	8.332	8.318	69.303	75
80	8 658.5	.00012	.00001	.1200	72 145.7	8.332	8.324	69.359	80
85	15 259.2	.00007	.00001	.1200	127 151.7	8.333	8.328	69.393	85
90	26 891.9	.00004		.1200	224 091.1	8.333	8.330	69.414	90
95	47 392.8	.00002		.1200	394 931.4	8.333	8.331	69.426	95
100	83 522.3	.00001		.1200	696 010.5	8.333	8.332	69.434	100

15% — Compound Interest Factors

| | Single Payment | | Uniform Payment Series | | | | Arithmetic Gradient | | |
| | Compound Amount Factor | Present Worth Factor | Sinking Fund Factor | Capital Recovery Factor | Compound Amount Factor | Present Worth Factor | Gradient Uniform Series | Gradient Present Worth | |
n	Find F Given P F/P	Find P Given F P/F	Find A Given F A/F	Find A Given P A/P	Find F Given A F/A	Find P Given A P/A	Find A Given G A/G	Find P Given G P/G	n
1	1.150	.8696	1.0000	1.1500	1.000	0.870	0	0	1
2	1.322	.7561	.4651	.6151	2.150	1.626	0.465	0.756	2
3	1.521	.6575	.2880	.4380	3.472	2.283	0.907	2.071	3
4	1.749	.5718	.2003	.3503	4.993	2.855	1.326	3.786	4
5	2.011	.4972	.1483	.2983	6.742	3.352	1.723	5.775	5
6	2.313	.4323	.1142	.2642	8.754	3.784	2.097	7.937	6
7	2.660	.3759	.0904	.2404	11.067	4.160	2.450	10.192	7
8	3.059	.3269	.0729	.2229	13.727	4.487	2.781	12.481	8
9	3.518	.2843	.0596	.2096	16.786	4.772	3.092	14.755	9
10	4.046	.2472	.0493	.1993	20.304	5.019	3.383	16.979	10
11	4.652	.2149	.0411	.1911	24.349	5.234	3.655	19.129	11
12	5.350	.1869	.0345	.1845	29.002	5.421	3.908	21.185	12
13	6.153	.1625	.0291	.1791	34.352	5.583	4.144	23.135	13
14	7.076	.1413	.0247	.1747	40.505	5.724	4.362	24.972	14
15	8.137	.1229	.0210	.1710	47.580	5.847	4.565	26.693	15
16	9.358	.1069	.0179	.1679	55.717	5.954	4.752	28.296	16
17	10.761	.0929	.0154	.1654	65.075	6.047	4.925	29.783	17
18	12.375	.0808	.0132	.1632	75.836	6.128	5.084	31.156	18
19	14.232	.0703	.0113	.1613	88.212	6.198	5.231	32.421	19
20	16.367	.0611	.00976	.1598	102.444	6.259	5.365	33.582	20
21	18.822	.0531	.00842	.1584	118.810	6.312	5.488	34.645	21
22	21.645	.0462	.00727	.1573	137.632	6.359	5.601	35.615	22
23	24.891	.0402	.00628	.1563	159.276	6.399	5.704	36.499	23
24	28.625	.0349	.00543	.1554	184.168	6.434	5.798	37.302	24
25	32.919	.0304	.00470	.1547	212.793	6.464	5.883	38.031	25
26	37.857	.0264	.00407	.1541	245.712	6.491	5.961	38.692	26
27	43.535	.0230	.00353	.1535	283.569	6.514	6.032	39.289	27
28	50.066	.0200	.00306	.1531	327.104	6.534	6.096	39.828	28
29	57.575	.0174	.00265	.1527	377.170	6.551	6.154	40.315	29
30	66.212	.0151	.00230	.1523	434.745	6.566	6.207	40.753	30
31	76.144	.0131	.00200	.1520	500.957	6.579	6.254	41.147	31
32	87.565	.0114	.00173	.1517	577.100	6.591	6.297	41.501	32
33	100.700	.00993	.00150	.1515	664.666	6.600	6.336	41.818	33
34	115.805	.00864	.00130	.1513	765.365	6.609	6.371	42.103	34
35	133.176	.00751	.00113	.1511	881.170	6.617	6.402	42.359	35
40	267.864	.00373	.00056	.1506	1 779.1	6.642	6.517	43.283	40
45	538.769	.00186	.00028	.1503	3 585.1	6.654	6.583	43.805	45
50	1 083.7	.00092	.00014	.1501	7 217.7	6.661	6.620	44.096	50
55	2 179.6	.00046	.00007	.1501	14 524.1	6.664	6.641	44.256	55
60	4 384.0	.00023	.00003	.1500	29 220.0	6.665	6.653	44.343	60
65	8 817.8	.00011	.00002	.1500	58 778.6	6.666	6.659	44.390	65
70	17 735.7	.00006	.00001	.1500	118 231.5	6.666	6.663	44.416	70
75	35 672.9	.00003		.1500	237 812.5	6.666	6.665	44.429	75
80	71 750.9	.00001		.1500	478 332.6	6.667	6.666	44.436	80
85	144 316.7	.00001		.1500	962 104.4	6.667	6.666	44.440	85

18% — Compound Interest Factors

| | Single Payment | | Uniform Payment Series | | | | Arithmetic Gradient | | |
| | Compound Amount Factor | Present Worth Factor | Sinking Fund Factor | Capital Recovery Factor | Compound Amount Factor | Present Worth Factor | Gradient Uniform Series | Gradient Present Worth | |
n	Find F Given P F/P	Find P Given F P/F	Find A Given F A/F	Find A Given P A/P	Find F Given A F/A	Find P Given A P/A	Find A Given G A/G	Find P Given G P/G	n
1	1.180	.8475	1.0000	1.1800	1.000	0.847	0	0	1
2	1.392	.7182	.4587	.6387	2.180	1.566	0.459	0.718	2
3	1.643	.6086	.2799	.4599	3.572	2.174	0.890	1.935	3
4	1.939	.5158	.1917	.3717	5.215	2.690	1.295	3.483	4
5	2.288	.4371	.1398	.3198	7.154	3.127	1.673	5.231	5
6	2.700	.3704	.1059	.2859	9.442	3.498	2.025	7.083	6
7	3.185	.3139	.0824	.2624	12.142	3.812	2.353	8.967	7
8	3.759	.2660	.0652	.2452	15.327	4.078	2.656	10.829	8
9	4.435	.2255	.0524	.2324	19.086	4.303	2.936	12.633	9
10	5.234	.1911	.0425	.2225	23.521	4.494	3.194	14.352	10
11	6.176	.1619	.0348	.2148	28.755	4.656	3.430	15.972	11
12	7.288	.1372	.0286	.2086	34.931	4.793	3.647	17.481	12
13	8.599	.1163	.0237	.2037	42.219	4.910	3.845	18.877	13
14	10.147	.0985	.0197	.1997	50.818	5.008	4.025	20.158	14
15	11.974	.0835	.0164	.1964	60.965	5.092	4.189	21.327	15
16	14.129	.0708	.0137	.1937	72.939	5.162	4.337	22.389	16
17	16.672	.0600	.0115	.1915	87.068	5.222	4.471	23.348	17
18	19.673	.0508	.00964	.1896	103.740	5.273	4.592	24.212	18
19	23.214	.0431	.00810	.1881	123.413	5.316	4.700	24.988	19
20	27.393	.0365	.00682	.1868	146.628	5.353	4.798	25.681	20
21	32.324	.0309	.00575	.1857	174.021	5.384	4.885	26.300	21
22	38.142	.0262	.00485	.1848	206.345	5.410	4.963	26.851	22
23	45.008	.0222	.00409	.1841	244.487	5.432	5.033	27.339	23
24	53.109	.0188	.00345	.1835	289.494	5.451	5.095	27.772	24
25	62.669	.0160	.00292	.1829	342.603	5.467	5.150	28.155	25
26	73.949	.0135	.00247	.1825	405.272	5.480	5.199	28.494	26
27	87.260	.0115	.00209	.1821	479.221	5.492	5.243	28.791	27
28	102.966	.00971	.00177	.1818	566.480	5.502	5.281	29.054	28
29	121.500	.00823	.00149	.1815	669.447	5.510	5.315	29.284	29
30	143.370	.00697	.00126	.1813	790.947	5.517	5.345	29.486	30
31	169.177	.00591	.00107	.1811	934.317	5.523	5.371	29.664	31
32	199.629	.00501	.00091	.1809	1 103.5	5.528	5.394	29.819	32
33	235.562	.00425	.00077	.1808	1 303.1	5.532	5.415	29.955	33
34	277.963	.00360	.00065	.1806	1 538.7	5.536	5.433	30.074	34
35	327.997	.00305	.00055	.1806	1 816.6	5.539	5.449	30.177	35
40	750.377	.00133	.00024	.1802	4 163.2	5.548	5.502	30.527	40
45	1 716.7	.00058	.00010	.1801	9 531.6	5.552	5.529	30.701	45
50	3 927.3	.00025	.00005	.1800	21 813.0	5.554	5.543	30.786	50
55	8 984.8	.00011	.00002	.1800	49 910.1	5.555	5.549	30.827	55
60	20 555.1	.00005	.00001	.1800	114 189.4	5.555	5.553	30.846	60
65	47 025.1	.00002		.1800	261 244.7	5.555	5.554	30.856	65
70	107 581.9	.00001		.1800	597 671.7	5.556	5.555	30.860	70

20% Compound Interest Factors

	Single Payment		Uniform Payment Series				Arithmetic Gradient		
	Compound Amount Factor	Present Worth Factor	Sinking Fund Factor	Capital Recovery Factor	Compound Amount Factor	Present Worth Factor	Gradient Uniform Series	Gradient Present Worth	
n	Find F Given P, F/P	Find P Given F, P/F	Find A Given F, A/F	Find A Given P, A/P	Find F Given A, F/A	Find P Given A, P/A	Find A Given G, A/G	Find P Given G, P/G	n
1	1.200	.8333	1.0000	1.2000	1.000	0.833	0	0	1
2	1.440	.6944	.4545	.6545	2.200	1.528	0.455	0.694	2
3	1.728	.5787	.2747	.4747	3.640	2.106	0.879	1.852	3
4	2.074	.4823	.1863	.3863	5.368	2.589	1.274	3.299	4
5	2.488	.4019	.1344	.3344	7.442	2.991	1.641	4.906	5
6	2.986	.3349	.1007	.3007	9.930	3.326	1.979	6.581	6
7	3.583	.2791	.0774	.2774	12.916	3.605	2.290	8.255	7
8	4.300	.2326	.0606	.2606	16.499	3.837	2.576	9.883	8
9	5.160	.1938	.0481	.2481	20.799	4.031	2.836	11.434	9
10	6.192	.1615	.0385	.2385	25.959	4.192	3.074	12.887	10
11	7.430	.1346	.0311	.2311	32.150	4.327	3.289	14.233	11
12	8.916	.1122	.0253	.2253	39.581	4.439	3.484	15.467	12
13	10.699	.0935	.0206	.2206	48.497	4.533	3.660	16.588	13
14	12.839	.0779	.0169	.2169	59.196	4.611	3.817	17.601	14
15	15.407	.0649	.0139	.2139	72.035	4.675	3.959	18.509	15
16	18.488	.0541	.0114	.2114	87.442	4.730	4.085	19.321	16
17	22.186	.0451	.00944	.2094	105.931	4.775	4.198	20.042	17
18	26.623	.0376	.00781	.2078	128.117	4.812	4.298	20.680	18
19	31.948	.0313	.00646	.2065	154.740	4.843	4.386	21.244	19
20	38.338	.0261	.00536	.2054	186.688	4.870	4.464	21.739	20
21	46.005	.0217	.00444	.2044	225.026	4.891	4.533	22.174	21
22	55.206	.0181	.00369	.2037	271.031	4.909	4.594	22.555	22
23	66.247	.0151	.00307	.2031	326.237	4.925	4.647	22.887	23
24	79.497	.0126	.00255	.2025	392.484	4.937	4.694	23.176	24
25	95.396	.0105	.00212	.2021	471.981	4.948	4.735	23.428	25
26	114.475	.00874	.00176	.2018	567.377	4.956	4.771	23.646	26
27	137.371	.00728	.00147	.2015	681.853	4.964	4.802	23.835	27
28	164.845	.00607	.00122	.2012	819.223	4.970	4.829	23.999	28
29	197.814	.00506	.00102	.2010	984.068	4.975	4.853	24.141	29
30	237.376	.00421	.00085	.2008	1 181.9	4.979	4.873	24.263	30
31	284.852	.00351	.00070	.2007	1 419.3	4.982	4.891	24.368	31
32	341.822	.00293	.00059	.2006	1 704.1	4.985	4.906	24.459	32
33	410.186	.00244	.00049	.2005	2 045.9	4.988	4.919	24.537	33
34	492.224	.00203	.00041	.2004	2 456.1	4.990	4.931	24.604	34
35	590.668	.00169	.00034	.2003	2 948.3	4.992	4.941	24.661	35
40	1 469.8	.00068	.00014	.2001	7 343.9	4.997	4.973	24.847	40
45	3 657.3	.00027	.00005	.2001	18 281.3	4.999	4.988	24.932	45
50	9 100.4	.00011	.00002	.2000	45 497.2	4.999	4.995	24.970	50
55	22 644.8	.00004	.00001	.2000	113 219.0	5.000	4.998	24.987	55
60	56 347.5	.00002		.2000	281 732.6	5.000	4.999	24.994	60

30% Compound Interest Factors

	Single Payment		Uniform Payment Series				Arithmetic Gradient		
	Compound Amount Factor	Present Worth Factor	Sinking Fund Factor	Capital Recovery Factor	Compound Amount Factor	Present Worth Factor	Gradient Uniform Series	Gradient Present Worth	
n	Find F Given P, F/P	Find P Given F, P/F	Find A Given F, A/F	Find A Given P, A/P	Find F Given A, F/A	Find P Given A, P/A	Find A Given G, A/G	Find P Given G, P/G	n
1	1.300	.7692	1.0000	1.3000	1.000	0.769	0	0	1
2	1.690	.5917	.4348	.7348	2.300	1.361	0.435	0.592	2
3	2.197	.4552	.2506	.5506	3.990	1.816	0.827	1.502	3
4	2.856	.3501	.1616	.4616	6.187	2.166	1.178	2.552	4
5	3.713	.2693	.1106	.4106	9.043	2.436	1.490	3.630	5
6	4.827	.2072	.0784	.3784	12.756	2.643	1.765	4.666	6
7	6.275	.1594	.0569	.3569	17.583	2.802	2.006	5.622	7
8	8.157	.1226	.0419	.3419	23.858	2.925	2.216	6.480	8
9	10.604	.0943	.0312	.3312	32.015	3.019	2.396	7.234	9
10	13.786	.0725	.0235	.3235	42.619	3.092	2.551	7.887	10
11	17.922	.0558	.0177	.3177	56.405	3.147	2.683	8.445	11
12	23.298	.0429	.0135	.3135	74.327	3.190	2.795	8.917	12
13	30.287	.0330	.0102	.3102	97.625	3.223	2.889	9.314	13
14	39.374	.0254	.00782	.3078	127.912	3.249	2.969	9.644	14
15	51.186	.0195	.00598	.3060	167.286	3.268	3.034	9.917	15
16	66.542	.0150	.00458	.3046	218.472	3.283	3.089	10.143	16
17	86.504	.0116	.00351	.3035	285.014	3.295	3.135	10.328	17
18	112.455	.00889	.00269	.3027	371.518	3.304	3.172	10.479	18
19	146.192	.00684	.00207	.3021	483.973	3.311	3.202	10.602	19
20	190.049	.00526	.00159	.3016	630.165	3.316	3.228	10.702	20
21	247.064	.00405	.00122	.3012	820.214	3.320	3.248	10.783	21
22	321.184	.00311	.00094	.3009	1 067.3	3.323	3.265	10.848	22
23	417.539	.00239	.00072	.3007	1 388.5	3.325	3.278	10.901	23
24	542.800	.00184	.00055	.3006	1 806.0	3.327	3.289	10.943	24
25	705.640	.00142	.00043	.3004	2 348.8	3.329	3.298	10.977	25
26	917.332	.00109	.00033	.3003	3 054.4	3.330	3.305	11.005	26
27	1 192.5	.00084	.00025	.3003	3 971.9	3.331	3.311	11.026	27
28	1 550.3	.00065	.00019	.3002	5 164.3	3.331	3.315	11.044	28
29	2 015.4	.00050	.00015	.3001	6 714.6	3.332	3.319	11.058	29
30	2 620.0	.00038	.00011	.3001	8 730.0	3.332	3.322	11.069	30
31	3 406.0	.00029	.00009	.3001	11 350.0	3.332	3.324	11.078	31
32	4 427.8	.00023	.00007	.3001	14 756.0	3.333	3.326	11.085	32
33	5 756.1	.00017	.00005	.3001	19 183.7	3.333	3.328	11.090	33
34	7 483.0	.00013	.00004	.3000	24 939.9	3.333	3.329	11.094	34
35	9 727.8	.00010	.00003	.3000	32 422.8	3.333	3.330	11.098	35
40	36 118.8	.00003	.00001	.3000	120 392.6	3.333	3.332	11.107	40
45	134 106.5	.00001		.3000	447 018.3	3.333	3.333	11.110	45

40% — Compound Interest Factors

n	Single Payment — Compound Amount Factor (Find F Given P) F/P	Single Payment — Present Worth Factor (Find P Given F) P/F	Uniform Payment Series — Sinking Fund Factor (Find A Given F) A/F	Uniform Payment Series — Capital Recovery Factor (Find A Given P) A/P	Uniform Payment Series — Compound Amount Factor (Find F Given A) F/A	Uniform Payment Series — Present Worth Factor (Find P Given A) P/A	Arithmetic Gradient — Gradient Uniform Series (Find A Given G) A/G	Arithmetic Gradient — Gradient Present Worth (Find P Given G) P/G
1	1.400	.7143	1.0000	1.4000	1.000	0.714	0	0
2	1.960	.5102	.4167	.8167	2.400	1.224	0.417	0.510
3	2.744	.3644	.2294	.6294	4.360	1.589	0.780	1.239
4	3.842	.2603	.1408	.5408	7.104	1.849	1.092	2.020
5	5.378	.1859	.0914	.4914	10.946	2.035	1.358	2.764
6	7.530	.1328	.0613	.4613	16.324	2.168	1.581	3.428
7	10.541	.0949	.0419	.4419	23.853	2.263	1.766	3.997
8	14.758	.0678	.0291	.4291	34.395	2.331	1.919	4.471
9	20.661	.0484	.0203	.4203	49.153	2.379	2.042	4.858
10	28.925	.0346	.0143	.4143	69.814	2.414	2.142	5.170
11	40.496	.0247	.0101	.4101	98.739	2.438	2.221	5.417
12	56.694	.0176	.00718	.4072	139.235	2.456	2.285	5.611
13	79.371	.0126	.00510	.4051	195.929	2.469	2.334	5.762
14	111.120	.00900	.00363	.4036	275.300	2.478	2.373	5.879
15	155.568	.00643	.00259	.4026	386.420	2.484	2.403	5.969
16	217.795	.00459	.00185	.4018	541.988	2.489	2.426	6.038
17	304.913	.00328	.00132	.4013	759.783	2.492	2.444	6.090
18	426.879	.00234	.00094	.4009	1 064.7	2.494	2.458	6.130
19	597.630	.00167	.00067	.4007	1 419.6	2.496	2.468	6.160
20	836.682	.00120	.00048	.4005	2 089.2	2.497	2.476	6.183
21	1 171.4	.00085	.00034	.4003	2 925.9	2.498	2.482	6.200
22	1 639.9	.00061	.00024	.4002	4 097.2	2.498	2.487	6.213
23	2 295.9	.00044	.00017	.4002	5 737.1	2.499	2.490	6.222
24	3 214.2	.00031	.00012	.4001	8 033.0	2.499	2.493	6.229
25	4 499.9	.00022	.00009	.4001	11 247.2	2.499	2.494	6.235
26	6 299.8	.00016	.00006	.4001	15 747.1	2.500	2.496	6.239
27	8 819.8	.00011	.00005	.4000	22 046.9	2.500	2.497	6.242
28	12 347.7	.00008	.00003	.4000	30 866.7	2.500	2.498	6.244
29	17 286.7	.00006	.00002	.4000	43 214.3	2.500	2.498	6.245
30	24 201.4	.00004	.00002	.4000	60 501.0	2.500	2.499	6.247
31	33 882.0	.00003	.00001	.4000	84 702.5	2.500	2.499	6.248
32	47 434.8	.00002	.00001	.4000	118 584.4	2.500	2.499	6.248
33	66 408.7	.00002	.00001	.4000	166 019.2	2.500	2.500	6.249
34	92 972.1	.00001		.4000	232 427.9	2.500	2.500	6.249
35	130 161.0	.00001		.4000	325 400.0	2.500	2.500	6.249

50% — Compound Interest Factors

n	Single Payment — Compound Amount Factor (Find F Given P) F/P	Single Payment — Present Worth Factor (Find P Given F) P/F	Uniform Payment Series — Sinking Fund Factor (Find A Given F) A/F	Uniform Payment Series — Capital Recovery Factor (Find A Given P) A/P	Uniform Payment Series — Compound Amount Factor (Find F Given A) F/A	Uniform Payment Series — Present Worth Factor (Find P Given A) P/A	Arithmetic Gradient — Gradient Uniform Series (Find A Given G) A/G	Arithmetic Gradient — Gradient Present Worth (Find P Given G) P/G
1	1.500	.6667	1.0000	1.5000	1.000	0.667	0	0
2	2.250	.4444	.4000	.9000	2.500	1.111	0.400	0.444
3	3.375	.2963	.2105	.7105	4.750	1.407	0.737	1.037
4	5.063	.1975	.1231	.6231	8.125	1.605	1.015	1.630
5	7.594	.1317	.0758	.5758	13.188	1.737	1.242	2.156
6	11.391	.0878	.0481	.5481	20.781	1.824	1.423	2.595
7	17.086	.0585	.0311	.5311	32.172	1.883	1.565	2.947
8	25.629	.0390	.0203	.5203	49.258	1.922	1.675	3.220
9	38.443	.0260	.0134	.5134	74.887	1.948	1.760	3.428
10	57.665	.0173	.00882	.5088	113.330	1.965	1.824	3.584
11	86.498	.0116	.00585	.5058	170.995	1.977	1.871	3.699
12	129.746	.00771	.00388	.5039	257.493	1.985	1.907	3.784
13	194.620	.00514	.00258	.5026	387.239	1.990	1.933	3.846
14	291.929	.00343	.00172	.5017	581.859	1.993	1.952	3.890
15	437.894	.00228	.00114	.5011	873.788	1.995	1.966	3.922
16	656.841	.00152	.00076	.5008	1 311.7	1.997	1.976	3.945
17	985.261	.00101	.00051	.5005	1 968.5	1.998	1.983	3.961
18	1 477.9	.00068	.00034	.5003	2 953.8	1.999	1.988	3.973
19	2 216.8	.00045	.00023	.5002	4 431.7	1.999	1.991	3.981
20	3 325.3	.00030	.00015	.5002	6 648.5	1.999	1.994	3.987
21	4 987.9	.00020	.00010	.5001	9 973.8	2.000	1.996	3.991
22	7 481.8	.00013	.00007	.5001	14 961.7	2.000	1.997	3.994
23	11 222.7	.00009	.00004	.5000	22 443.5	2.000	1.998	3.996
24	16 834.1	.00006	.00003	.5000	33 666.2	2.000	1.999	3.997
25	25 251.2	.00004	.00002	.5000	50 500.3	2.000	1.999	3.998
26	37 876.8	.00003	.00001	.5000	75 751.5	2.000	1.999	3.999
27	56 815.1	.00002	.00001	.5000	113 628.3	2.000	2.000	3.999
28	85 222.7	.00001		.5000	170 443.4	2.000	2.000	3.999
29	127 834.0	.00001		.5000	255 666.1	2.000	2.000	3.999
30	191 751.1	.00001		.5000	383 500.1	2.000	2.000	4.000
31	287 626.6			.5000	575 251.2	2.000	2.000	4.000
32	431 439.9			.5000	862 877.8	2.000	2.000	4.000

State Boards of Registration

ALABAMA
State Board of Registration for
Professional Engineers and Surveyors
300 Interstate Park Drive, Suite 301
Montgomery 36130

ALASKA
State Board of Registration for
Architects, Engineers and Land
Surveyors
State Office Building, 9th Floor
P.O. Box D-LIC
Juneau 99811

ARIZONA
State Board of Technical Registration
5060 N. 19th Avenue, Suite 306
Phoenix 85015

ARKANSAS
State Board of Registration for
Professional Engineers and Land
Surveyors
Twin City Bank Building, Suite 700
P.O. Box 2541
Little Rock 72203

CALIFORNIA
Board of Registration for Professional
Engineers and Land Surveyors
1428 Howe Avenue, Suite 56
Sacramento 95825-3298

COLORADO
State Board of Registration for
Professional Engineers and
Professional Land Surveyors
600-B State Services Building
1525 Sherman Street
Denver 80203

CONNECTICUT
State Board of Examiners for
Professional Engineers and Land
Surveyors
State Office Building, Room G-3A
165 Capitol Avenue
Hartford 06106

DELAWARE
Delaware Association of Professional
Engineers
2005 Concord Pike
Wilmington 19803

DISTRICT OF COLUMBIA
Board of Registration for Professional
Engineers
614 H Street, N.E., Room 923
Washington 20001

FLORIDA
Board of Professional Engineers
130 North Monroe Street
Tallahassee 32399-0750

GEORGIA
State Board of Registration for
Professional Engineers and Land
Surveyors
166 Prior Street, S.W.
Atlanta 30303

GUAM
Territorial Board of Registration for
Professional Engineers, Architects and
Land Surveyors
Guam Department of Public Works
P.O. Box 2950
Agana 96910

HAWAII
Board of Registration for Professional
Engineers, Architects, Surveyors, and
Landscape Architects
1010 Richards Street
P.O. Box 3469
Honolulu 96801

IDAHO
Board of Professional Engineers and
Land Surveyors
842 La Cassia Drive
Boise 83705

ILLINOIS
Department of Professional
Regulation
Professional Engineers' Examining
Committee
320 West Washington Street, 3rd Floor
Springfield 62786

INDIANA
State Board of Registration for
Professional Engineers and Land
Surveyors
1021 State Office Building
100 North Senate Avenue
Indianapolis 46204

IOWA
Engineering & Land Surveying
Examining Board
1918 S.E. Hulsizer
Ankeny 50021

KANSAS
State Board of Technical Professions
Landon State Office Building
900 Jackson, Suite 507
Topeka 66612-1214

KENTUCKY
State Board of Registration for
Professional Engineers and Land
Surveyors
Kentucky Engineering Center
160 Democrat Drive
Frankfort 40601

LOUISIANA
State Board of Registration for
Professional Engineers and Land
Surveyors
1055 St. Charles Avenue, Suite 415
New Orleans 70130-3997

MAINE
State Board of Registration for
Professional Engineers
State House, Station 92
Augusta 04333

MARYLAND
State Board of Registration for
Professional Engineers
501 St. Paul Place, Room 902
Baltimore 21202

MASSACHUSETTS
Board of Registration for Professional
Engineers and Land Surveyors
Saltonstall Building, Room 1512
100 Cambridge Street
Boston 02202

MICHIGAN
Board of Professional Engineers
611 West Ottawa
P.O. Box 30018
Lansing 48909

MINNESOTA
State Board of Registration for
Architects, Engineers, Land Surveyors
and Landscape Architects
402 Metro Square
121 East 7th Street
St. Paul 55101

MISSISSIPPI
State Board of Registration for
Professional Engineers and Land
Surveyors
239 N. Lamar, Suite 501
P.O. Box 3
Jackson 39205

MISSOURI
Board for Architects, Professional
Engineers and Land Surveyors
3523 North Ten Mile Drive
P.O. Box 184
Jefferson City 65102

MONTANA
Board of Professional Engineers and
Land Surveyors
1424 9th Avenue
Helena 59620-0407

NEBRASKA
State Board of Examiners for
Professional Engineers and Architects
301 Centennial Mall, South
P.O. Box 94751
Lincoln 68509

NEVADA
State Board of Registered
Professional Engineers and Land
Surveyors
1755 East Plumb Lane, Suite 135
Reno 89502

NEW HAMPSHIRE
Board of Professional Engineers
Storrs Street
Concord 03301

NEW JERSEY
State Board of Professional Engineers
and Land Surveyors
1100 Raymond Blvd, Room 317
Newark 07102

NEW MEXICO
Board of Registration for Professional
Engineers and Surveyors
440 Cerrillos Road
Suite A
Santa Fe 87501

NEW YORK
State Board for Engineering and Land
Surveying
State Education Dept., Cultural
Education Center
Madison Avenue
Albany 12230

NORTH CAROLINA
Board of Registration for Professional
Engineers and Land Surveyors
3620 Six Forks Road
Raleigh 27609

NORTH DAKOTA
State Board of Registration for
Professional Engineers and Land
Surveyors
420 Avenue B East
P.O. Box 1357
Bismarck 58502

NORTHERN MARIANA ISLANDS
Board of Professional Licensing
P.O. Box 449 CHRB
Siapan, MP 96950

OHIO
State Board of Registration for
Professional Engineers and Surveyors
77 South High Street, 16th Floor
Columbus 43266-0314

OKLAHOMA
State Board of Registration for
Professional Engineers
and Land Surveyors
Oklahoma Engineering Center,
Room 120
201 N.E. 27th Street
Oklahoma City 73105

OREGON
Board of Engineering Examiners
750 Front Street, N.E., Suite 240
Salem 97310

PENNSYLVANIA
State Registration Board for
Professional Engineers
613 Transportation & Safety Building
P.O. Box 2649
Harrisburg 17105-2649

PUERTO RICO
Board of Examiners of Engineers,
Architects and Land Surveyors
Ciputacion Provencial Building
Room 201
P.O. Box 3271
San Juan 00904

RHODE ISLAND
State Board of Registration for
Professional Engineers and Land
Surveyors
233 Richmond Street, Suite 230
Providence 02903

SOUTH CAROLINA
State Board of Registration for
Professional Engineers and Land
Surveyors
2221 Devine Street, Suite 404
P.O. Drawer 50408
Columbia 29250

SOUTH DAKOTA
State Commission of Engineering and
Architectural Examiners
2040 West Main Street, Suite 304
Rapid City 57702-2497

TENNESSEE
State Board of Architectural and
Engineering Examiners
500 James Robertson Parkway
3rd Floor
Nashville 37219

TEXAS
State Board of Registration for
Professional Engineers
1917 IH 35 South
P.O. Drawer 18329
Austin 78760

UTAH
Representative Committee for
Professional Engineers and Land
Surveyors
160 East 300 South
P.O. Box 45802
Salt Lake City 84145

VERMONT
Board of Registration for Professional
Engineers
Pavilion Building
Montpelier 05602

VIRGINIA
Board for Architects, Professional
Engineers, Land Surveyors and
Landscape Architects
3600 West Broad Street
Richmond 23230-4917

VIRGIN ISLANDS
Board for Architects, Engineers and
Land Surveyors
Property and Procurement Building
No. 1 Sub Base, Room 205
St. Thomas 00801

WASHINGTON
State Board of Registration for
Professional Engineers and Land
Surveyors
1300 Quince Street, 4th Floor
P.O. Box 9649
Olympia 98504

WEST VIRGINIA
State Board of Registration for
Professional Engineers
608 Union Building
Charleston 25301

WISCONSIN
Examining Board of Architects,
Professional Engineers, Designers
and Land Surveyors
1400 East Washington Avenue
P.O. Box 8935
Madison 53708-8935

WYOMING
State Board of Registration for
Professional Engineers and
Professional Land Surveyors
Herschler Building, Room 4135 E
Cheyenne 82002

Index